Hardware Technologies for Artificial Intelligence

In this comprehensive reference work for researchers, engineers, and students, Kawahara provides a one-stop exploration of next-generation computing at the large-scale integration (LSI) circuit level, with a focus on the integration of artificial intelligence (AI), advanced LSI design, Ising machines, and memory innovations.

While current GPUs have high parallel processing capabilities suitable for computations on large datasets, their power consumption is approaching its limit and requires further development. Additionally, edge computing is becoming increasingly important alongside cloud computing. Amid these significant technological trends, this book provides readers with important insights into next-generation computing, namely (1) neural network (AI) LSIs and their low power and high performance, (2) hardware design technology for combinatorial optimization problems and Ising machines, and (3) semiconductor memory and data-centric computing. Kawahara first describes the basics of LSI design and neural networks and then describes their large-scale integration, power efficiency, and performance enhancements. Then, he also explains hardware design techniques for Ising machines and offers case studies of fully coupled Ising machine LSI. Lastly, he discusses the basics of semiconductor memory and near/in-memory AI computing, and then, he examines the future prospects. Readers will be able to apply this knowledge to the design and manufacture of such devices to overcome the limitations of current hardware and computational methods, driving future advancements in AI and optimization.

This is a valuable reference for students, engineers, and researchers alike in this field. As it begins with the basics, it enables all readers to follow the direction of next-generation computing and its important technical content without the need for prior knowledge or reference to other books.

Takayuki Kawahara is a professor at Tokyo University of Science. He earned his bachelor's, master's, and doctorate degrees from Kyushu University in 1983, 1985, and 1993, respectively. He has significant experience within both industry and academia. He is a member of the Institute of Electronics, Information and Communication Engineers (IEICE) and a fellow of the Institute of Electrical and Electronics Engineers (IEEE).

Hardware Technologies for Artificial Intelligence

AI Chips, Ising Machines, and In-Memory Computing

Takayuki Kawahara

CRC Press
Taylor & Francis Group
Boca Raton London New York

CRC Press is an imprint of the
Taylor & Francis Group, an **informa** business

Designed cover image: Shutterstock

First edition published 2026
by CRC Press
2385 NW Executive Center Drive, Suite 320, Boca Raton FL 33431

and by CRC Press
4 Park Square, Milton Park, Abingdon, Oxon, OX14 4RN

CRC Press is an imprint of Taylor & Francis Group, LLC

ISBN: 9781032985121 (hbk)
ISBN: 9781032985138 (pbk)
ISBN: 9781003599067 (ebk)

DOI: 10.1201/9781003599067

Typeset in Palatino
by codeMantra

Contents

Preface

In recent years, with the rapid spread of IoT (Internet of Things), sensors and actuators have been installed everywhere, and a social infrastructure is being built in which they are connected to the cloud via networks. In various fields such as manufacturing, transportation, energy management, and even medical care and nursing, IoT is developing as a key technology that enables real-time information collection and decision-making.

However, IoT generates massive amounts of data, and attempting to process all of them in the cloud can lead to network congestion, processing delays (latency), and privacy concerns. To address these challenges, edge computing, a distributed processing architecture, is gaining attention. By processing data where they are generated, edge computing reduces network traffic while enhancing real-time performance and security.

However, edge processing also poses challenges. It must be able to operate in environments where there is limited installation space and power. Therefore, dedicated hardware that efficiently performs "artificial intelligence" (AI) processing is becoming important. In particular, technologies that enable deep learning-based neural networks to run at high speeds and with low power consumption are attracting attention. Specialized AI chips (e.g., those built using wired logic or integrated logic and memory architectures) can achieve power performance ratios that are unattainable with conventional Central Processing Units (CPUs) and Graphics Processing Units (GPUs). Furthermore, with the recent spread of FPGAs (field programmable gate arrays), it is becoming possible to flexibly design and verify AI hardware mechanisms. This feature offers readers valuable insights into the latest advancements in AI hardware development.

This book discusses this technology from three perspectives:

1. Structure of AI chip circuits

   AI chip circuits are created by integrating neural networks into integrated circuits. Starting with a basic explanation of neural networks, this part provides an overview of the basic elemental circuits used in integrated circuits and then shows the structure of AI chips at the circuit block level. The structure is designed to enable parallel processing by incorporating a large number of processing units. Furthermore, this part describes technologies for reducing power consumption and improving the performance of these AI chips, which are important for edge computing.

2. Fully coupled Ising machines that solve combinatorial optimization problems

   At the edge, many problems require finding the optimal combination of multiple elements to maximize a certain metric. The fully coupled Ising machine, which solves this combinatorial optimization problem, has the same structure as the Hopfield network, which is a type of neural network. By incorporating an annealing mechanism into this structure, we obtain the fully coupled Ising machine. We also provide a detailed explanation of the LSI implementation of this fully connected Ising machine for use at the edge and present a concrete example that achieves high power performance.

3. Semiconductor memory and in-memory computing technology

   This part introduces near-memory computing and in-memory computing technologies, which are important for improving the performance and power efficiency of AI hardware at the edge. The power consumption issue in hardware is caused by data movement, but this can be solved by specializing the hardware for AI, making it a field with great potential for future development. This part begins with a basic explanation of semiconductor memory. It then explains the main types of memory and describes methods for applying them to AI hardware.

This book is an introductory guide to AI hardware technology, from the basics to applications, aimed at science and engineering students and young researchers interested in the intersection of AI, IoT, and integrated circuit design. Individuals interested in the hardware that supports the "brain" in the AI era will find this concise book useful.

Takayuki Kawahara

1

Introduction: AI Computing

1.1 Artificial Intelligence (AI) Processing

Artificial intelligence (AI) is defined as a technology and device that performs information processing similar to human thought processes or that humans perceive as intellectual and is considered a fundamental technology for industrial development. It is well known that the term "artificial intelligence" was first used in 1956 at a conference called the Dartmouth Conference [1]. Since the 1960s, attempts have been made to have computers perform inference, search [2], and machine translation, and development continued into the expert systems [3] of the 1980s.

On the other hand, research on neural networks (NNs) [4], which aim to artificially reproduce the neural circuits of the brain, has been inspired by studies of the physiology of the brain and nerves. The Perceptron [5] in 1958 and the Cognitron [6] in 1975 for pattern recognition, and the Hopfield model [7,8], which can handle combinatorial optimization problems, were developed. Methods for handling large networks, such as the error backpropagation method [9], were also developed.

Later, as the cost of data acquisition and that of computing power dropped significantly with the development of the Internet and computer technology, the 2010s led to the so-called deep learning [10,11] breakthroughs. Here, the fields of AI and NNs came to be treated as one and the same. Deep learning is a type of AI processing that uses multilayer NNs and is applied to tasks such as image recognition, speech recognition, and natural language processing based on huge amounts of data to perform advanced feature extraction and prediction. In the "information processing that humans perceive as intellectual," which is also the definition of AI, for example, image recognition has been reported to exceed human accuracy. In this book, information processing by this integrated AI/NN is called AI processing. As if in response to the development of AI processing, the Internet of Things (IoT) has progressed, and edge computing has become important. Edge computing is a technology that processes data from sensors and other sources at a location close to the site where the data are generated (the edge). In recent years, a huge amount of data has been generated at the edge, and as valuable information is created

DOI: 10.1201/9781003599067-1

from this vast amount of data, the importance of AI processing at the edge (edge AI) is increasing proportionally. By sending data to the cloud after AI processing at the edge, AI processing is also performed in the cloud, but the amount of communication and processing in the cloud can be reduced, and the processing speed and security of the overall system can be increased.

The realization of this technology requires smart processing of huge amounts of data, and the technology of AI processing integrated circuit (AI chip) is being further developed to achieve higher computational processing power and a higher power-to-performance ratio through integrated circuit technology (LSI technology). This AI chip will be a chip that performs specialized processing for specific applications.

1.2 Artificial Intelligence and Deep Learning

Let us take a closer look at current AI technology using Figure 1.1a. As mentioned above, AI technology is the technology that performs information processing that humans perceive as intellectual. Historically, AI began with rule-based systems and expert systems. These systems used information registered by humans and performed processing mechanisms decided by humans at a level comparable to that of experts. Later came machine learning. This is a technology in which AI itself learns multiple rules/patterns based on a large amount of given data to classify and predict. First, k-nearest neighbor algorithm (KNN) and support vector machine (SVM) were developed. As mentioned in the previous section, NNs became important in the development of machine

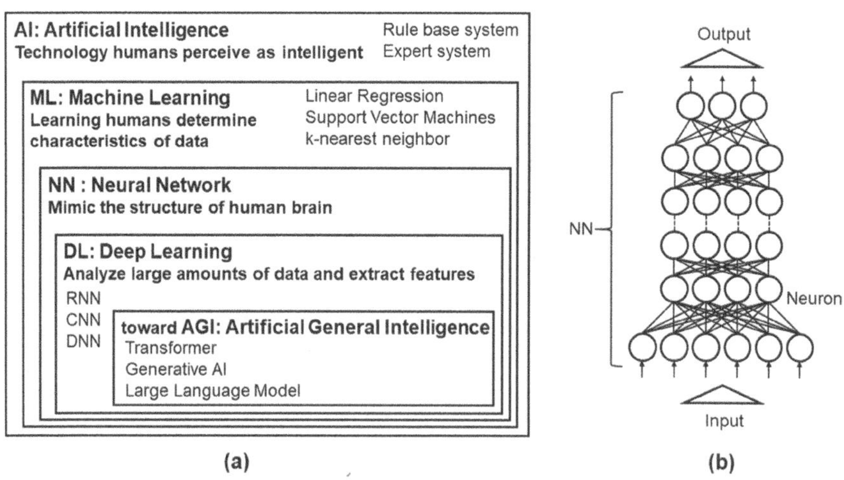

FIGURE 1.1
Artificial intelligence. (a) Artificial intelligence and (b) neural network.

learning. This network system has been engineered to emulate the NN of the brain, thereby enabling advanced machine learning. This is what we will be discussing in this book. Schematically, a NN is a system of many interconnected processing units called neurons, as shown in Figure 1.1b. As shown in the figure, multiple neurons form layers from input to output as an image, and machine learning in which information is processed by a NN with a multilayer structure is called deep learning. It has been found that this structure can be used to achieve advanced image recognition, speech recognition, translation, and more, and has since been further developed. Deep neural network (DNN), convolutional neural network (CNN), recurrent neural network (RNN), etc., are available. In addition, there is a growing expectation that artificial general intelligence (AGI), which is not human intelligence itself, but something comparable to it, can be created. Not yet at the AGI stage, transformer, generative AI, and LLM have been developed.

1.3 Learning and Inference

In current AI using NNs, the process is largely divided into learning and inference. In learning, NNs learn patterns and relationships in data using a large amount of data (training data set). In inference, the learned NNs are used to recognize, classify, and predict data not seen at the time of learning. The actual learning, which will be discussed in detail in later chapters, is the adjustment of the weights of the connections between neurons in one layer and neurons in the next layer of a NN, as well as the biases of each layer. This is how the functional relationship between inputs and outputs is obtained. This learning requires a large amount of data in advance and a long learning operation using this data, i.e., a lot of computational processing. Once this is obtained, the NN can, as an inference, produce corresponding outputs for different but similar data than at the time of learning.

There are the following types of learning. First, supervised learning is a method in which the model is trained on data that contains the correct answer (labeled data). However, after the model is trained in this way, the goal is to get the model to answer correctly on data that does not contain the correct answer. Next, unsupervised learning is a mechanism in which the machine algorithm extracts the essential structure and rules of the given data. Rather than labelling the correct answer data in advance, we allow the machine to learn so that the model can learn to capture the characteristics of the data. Furthermore, reinforcement learning is a mechanism whereby a machine learns to perform optimal behaviour through repeated trial and error without being given the correct answer. It is a learning process in which the machine is given a reward based on how good the behaviour was and is made to act in such a way that the reward is high.

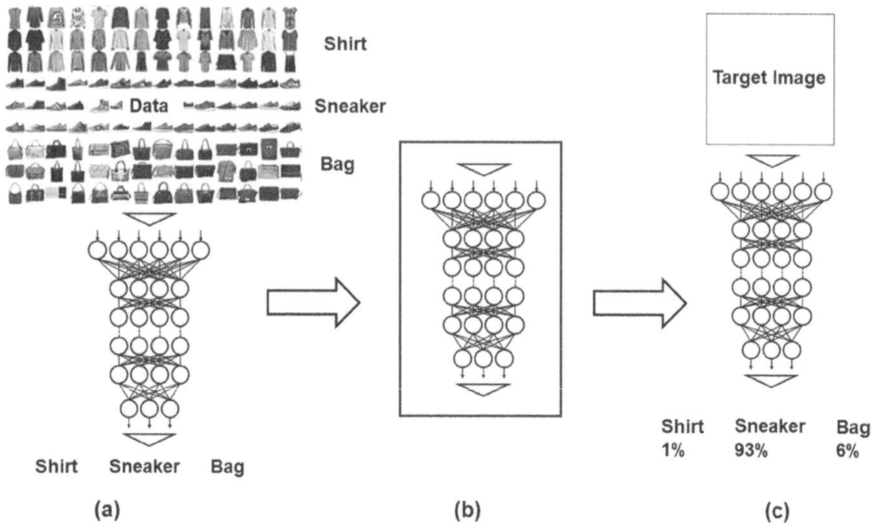

FIGURE 1.2
Learning and inference. (a) Learning, (b) trained network, and (c) inference.

Figure 1.2 shows an example. Here, the data shown in Figure 1.2a are multiple images labeled as shirts, sneakers, and bags. This is a dataset known as Fashion-mnist [12]. During training, the image data are given one by one to the input of the NN. The output is assumed to be able to produce a label for a shirt, a sneaker, or a bag. Then, for each single image, the output result is compared to see if it is the correct label for the given image, and the weights and biases between the layers are adjusted in the NN. This process is conducted for a substantial volume of image data, with the NN serving the function of guaranteeing that these correspond, that is to say, ensuring the correct correspondence between the input images and the output labels. This is the learning process. Consequently, the trained NN depicted in Figure 1.2b is obtained. Next, as an inference, an image as shown in Figure 1.2c is given to the input of this NN. This can be different from the data used during training. In contrast, the output produces results for three candidate labels, e.g., 1% shirt, 93% sneaker, and 6% bag. In this case, therefore, the item can be identified as a sneaker. This is the inference.

1.4 Integrated Circuits for Information Processing and AI Chips

Before describing this AI chip that enables AI processing, we first outline the differences from conventional integrated circuits for information processing.

As an integrated circuit that performs information processing, the central processing unit (CPU) [13] is the central part of a computer and is responsible for program processing, calculation (arithmetic processing), and control of computer components. High performance of a single CPU means that it can process and output input data with a certain bit width as quickly as possible, and its performance is determined by its bit width and operating frequency. However, the performance improvement of a single CPU has been slowing down since around 2005. This is due to power consumption issues and the slowdown in the progress of miniaturization represented by Moore's Law. On the other hand, unlike the performance of a single CPU, the volume of data communication has continued to increase over the years. This is, of course, due to the increase in data volume itself, which has also been referred to as the era of big data since around the 2010s. The communication environment has also improved, and many data centers have been built. In addition, with the development of IoT, a huge amount of data is being generated from various sensors and devices. In other words, while the performance of a single CPU is slowing down, the volume of data communication and the volume of data itself are continuing to grow.

Figure 1.3 shows a schematic of CPU and graphics processing unit (GPU) [13] configurations. In order to cope with the above-mentioned increase in data volume and its communication, the following studies have been conducted in these systems.

First, there are two types of CPU innovations to improve computing performance: multicore (Figure 1.3a) and many-core. A multicore is one that identifies the parts of a program that can be parallelized with respect to the processing described in the program and executes them using multiple processor cores. The core itself has a complete instruction set, and so

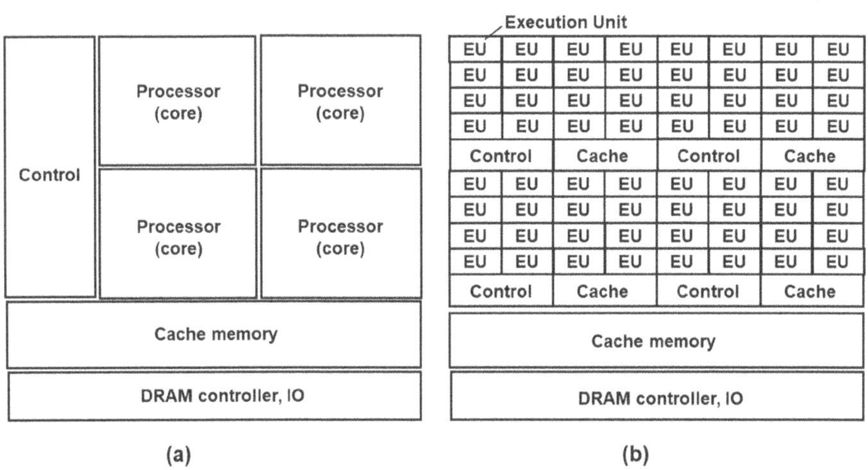

FIGURE 1.3
CPU and GPU. (a) CPU and (b) GPU.

far, the number of cores has roughly doubled with each design generation. Many-core systems aim to increase the execution throughput of the amount of data handled in the program portions that can be processed in parallel. It has many cores, each of which is highly multithreaded. This is intended to process more data efficiently.

One of the characteristics of this development in processing large amounts of data is that a large amount of data is prepared in advance and processed in batches. Here, it is important to be able to efficiently process a large amount of prepared data in a certain amount of time (throughput). The development of the GPU shown in Figure 1.3b has progressed in response to this data processing feature, albeit initially from a different application. GPUs are chips that specialize in graphics processing, and while they are faster than CPUs in terms of floating-point arithmetic, they are also faster due to the large amount of parallel processing. This is why GPUs are used beyond the gaming field for processing large amounts of data. CPUs and GPUs are used in a complementary manner, with CPUs controlling programs and managing data, and GPUs performing arithmetic operations on large amounts of data.

However, AI processing requires processing large amounts of data at once, efficiently, and at low power, and power is becoming an issue for GPUs, making it difficult to achieve higher performance. Therefore, there is a need to develop integrated circuits specialized for AI processing. The technology that can solve this problem is an AI chip, which is an integrated circuit specialized for AI processing itself.

1.5 AI Chip Overview

What is an integrated circuit specialized for AI processing? Let us examine some excerpts from this book.

First, since deep learning, the fields of AI and NNs have become one and the same. In this AI process, the device that performs this process consists of multilayered neurons (nodes). At this time, as shown in Figure 1.4a, each neuron weights the input signal and calculates the output using an activation function. NNs are known to be able to represent various functions depending on their structure. Here, a function is a relationship between inputs and outputs. In other words, by adjusting the weights between layers of NN and biases, the functional relationship between input and output can be obtained through learning.

What would be a suitable configuration of integrated circuits, or AI chips, for this realization? The purpose of this book is to understand this from the circuit level from three aspects (AI chips themselves, Ising machines, and in-memory computing), although the configuration of the unit block that performs the computation is described here. This is an array structure in which

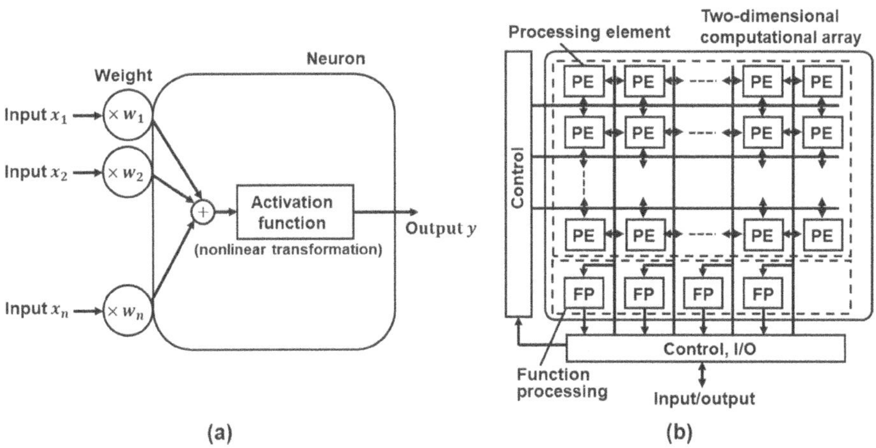

FIGURE 1.4
Neuron structure and AI chip structure. (a) Neuron structure and (b) main components of the AI chip.

multiple types of processing elements (in this case, two types) are laid out in two dimensions, as shown in Figure 1.4b. Here, the first step is to realize an arithmetic process that performs weighting on the input signal. The elements of this arithmetic processing consist of a circuit that stores the weights and a circuit that performs the operation. Furthermore, for each row of this two-dimensional array, there is an element of function processing that performs processing of the activation function. This structure enables the processing shown in Figure 1.4a. In this case, since many small arithmetic units are built in, it is possible to process many operations in parallel at the same time. Compared to a CPU, the AI chip has several to several hundred times the computing speed, consumes less power, and has higher computing performance when executing the same operations.

The above contents will be studied in this book. It has also been shown that seeking solutions to a problem with wide applications in everyday life and industry, called the combinatorial optimization problem, can be solved in a similar way to what is called a Hopfield network [8]. In both cases, the configuration is suitable for preparing a large amount of data in advance and processing it in batches.

However, AI chips are specialized for specific arithmetic operations and are not suited for general-purpose processing. In addition, the use of AI chips requires optimization of AI algorithms, and design and development require an understanding of these algorithms. Therefore, the use of AI chips is suitable when high-speed arithmetic processing is required, specifically for a particular application or when a large amount of data needs to be handled. Conversely, CPUs are suitable for general applications, and careful consideration should be given to whether there are benefits to be gained from the use of AI chips.

1.6 Edge AI

In this section, we discuss processing at the edge, which is important for AI chip applications. Processing at the edge is the processing of data from devices, sensors, etc., close to the site where it occurs (the edge). This processing is computing using limited computing resources at the location where the data is located. Processing in the cloud, on the other hand, is computing that can utilize a large amount of computing resources without worrying about the physical location of the data via the Internet.

A comparison of processing in the cloud and at the edge in the general case is summarized in Table 1.1. The processing is characterized by large amounts of data processing and advanced calculations in the cloud, while real-time data processing without delay is performed at the edge, close to the data generation. In addition, although there are significant differences depending on the application and purpose, memory in the cloud can use DDR4 memory [14] or high bandwidth memory (HBM) [15], with capacities ranging from tens to hundreds of GB, while memory in the edge uses LPDDR4 memory [16] or embedded MultiMediaCard (eMMC) [17], with only a few GB of capacity available. In terms of power, the cloud can use dedicated power and cooling systems, and the power consumption of the AI processing part alone can reach several tens of kilowatts per rack. On the other hand, since battery operation and power saving are required at the edge, the power for AI

TABLE 1.1
Comparison of Information Processing in the Cloud and on the Edge

	Cloud/Server	Edge
Image		
Processing features	Massive data processing, advanced calculations	Real-time data processing
Memory	TB, DRAM, Hybrid NVDRAM	GB, SRAM, NV-RAM
AI (artificial intelligence) processing	learning, inference	Inference, control
AI processing power	Tens of kW/rack, exclusive power supply	Several mW (sensor embedded) to several W (board)
Implementation environment	Air-conditioned room, dedicated rack	Factory, engine room, wind, and rain environments

processing is less than a few watts, and even AI processing may require mW level when placed near sensors. In terms of implementation, the cloud offers the flexibility to build a rack with many processors and memory, so there are relatively few physical restrictions. Comparing the performance aspects of AI processing performed by the two systems, the cloud enables complex processing by utilizing large-scale, high-performance resources. However, there are delays in data transmission and response times. On the other hand, at the edge, real-time performance is often required, requiring fast and low-latency response with limited resources. AI processing includes learning, inference, and control, which will be explained in a later chapter. Learning and inference are performed in the cloud, while inference and control are mainly performed at the edge.

The reasons for developing AI LSI at the edge are as follows.

1. It reduces latency, defined as the time required for data communication, when sending data to the cloud.
2. This approach conserves bandwidth. Transmitting substantial volumes of data to the cloud can pose challenges due to bandwidth constraints.
3. It ensures security. Transmitting data to the cloud introduces the potential risk of confidential information being compromised.

At the edge, there are greater limitations on the area where it can be implemented and the power it can use. Therefore, a compact and high-performance AI LSI is needed.

As the IoT develops further, more complex data will be required for processing at the edge. As shown in Figure 1.5, in the conventional IoT of Figure 1.5a, the data at the edge were temperature, CO_2 concentration, and current of equipment, etc. In the future IoT of Figure 1.5b, data such as images, audio, and video will become more important to be processed at the edge. For example, in a smart home environmental monitoring system, conventional edge processing processes data obtained from temperature and CO_2 sensors to monitor room temperature and indoor air quality. However, in the future, edge devices will have sensors with cameras and microphones, enabling environmental monitoring through images, audio, and video. This

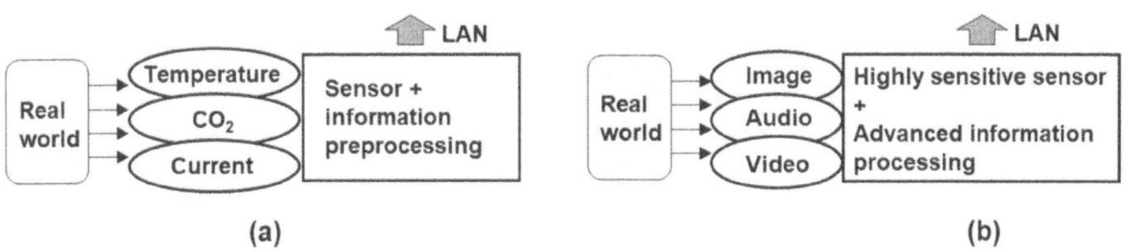

FIGURE 1.5
IoT development. (a) Conventional IoT and (b) next-generation IoT.

data can then be processed at the edge to detect the presence and activity of people and abnormal sound and video patterns. This can provide a comfortable living environment, help optimize lifestyles, or provide a high level of security and safety similar to that of home security service. It should also be noted that this includes the value of improving user experience and quality of life. In addition to processing data such as engine speed and engine revolutions, the future of automotive applications will require real-time processing of video and object detection data obtained from cameras and sensor data from inside and outside the vehicle at the edge. It is said that this will enable advanced driving assistance such as recognition of obstacles and signals on the road, automatic braking, and lane monitoring.

Thus, data at the edge will increase in both quality and quantity, requiring more sophisticated processing and decision-making. Processing at the edge is expected to enable smarter and more efficient systems to keep pace with the evolution of the IoT.

AI chips will be required to perform these tasks at the edge, but in conjunction with the cloud. This will also be important for the application of processing in the cloud itself.

1.7 AI Chips Pioneer the Future

The development of AI chip technology, which you will learn about in this book, is expected to lead to sustainable development in areas such as those shown in Figure 1.6, as data processing becomes smarter.

Residential and building: It will be possible to automate energy conservation and security management in homes and buildings (smart homes and smart buildings). AI chips embedded in sensors can incorporate functions to meticulously control lighting and air conditioning in living rooms, thereby reducing power waste. This can also provide a comfortable living environment, help optimize lifestyles, or provide a high level of security and safety.

Automated driving and transportation systems: High-speed image recognition and object detection at the edge and data analysis using deep learning are expected to advance the realization of automated driving vehicles. Advanced on-the-fly analysis is essential for self-driving cars, and cars equipped with AI chips will be able to drive automatically without the need for the driver to operate the car. This is expected to reduce traffic accidents and improve traffic efficiency. The use of AI in the optimization of the entire transportation system will also help to reduce traffic congestion and improve the convenience of public transportation.

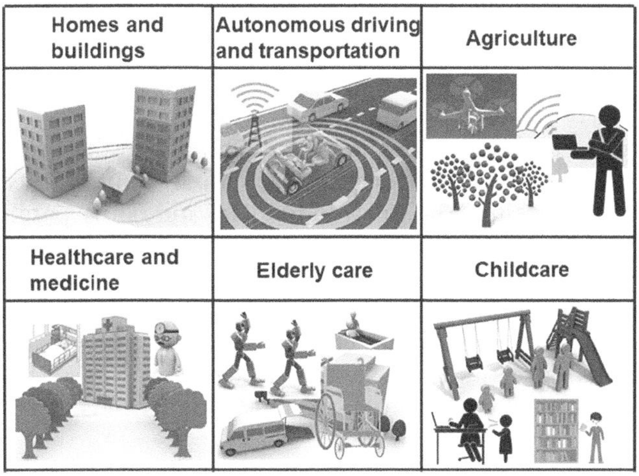

FIGURE 1.6
Examples of areas where AI chips can be used.

Agriculture: Artificial intelligence technology is expected to automate farm work and improve production. AI chips will be widely used here as smart agriculture. For example, sensors could be installed in fields to acquire real-time information on soil conditions and weather conditions, enabling not only automatic watering and fertilizer application but also individual care by detecting the condition of the crops themselves. In the near future, the integration of robotics will enable farmers to increase crop yields without the need for human labor.

Healthcare and medicine: AI chips can be integrated into medical devices and wearable devices. AI chips could be used, for example, to monitor patients' vital data in real time and detect abnormalities at an early stage, enabling prevention and early detection of diseases. AI chips will be used to analyze medical images and assist in diagnosis, reducing the burden on physicians and enabling them to provide high-quality care to more patients. AI technology itself will be able to suggest optimal treatments based on a patient's genetic information and lifestyle habits, which is expected to lead to more effective treatments.

Elderly care: AI chips would be used to monitor the health status and behavior patterns of the elderly; sensors and wearable devices equipped with AI chips could constantly monitor the heart rate and activity levels of the elderly and notify family members and caregivers of any abnormalities, thereby enabling rapid response. In addition, although combined with robot technology, AI-based robots can facilitate communication with the elderly, reduce the burden on caregivers, and support mental health as well. As a further possibility, robots equipped with AI chips can help with household chores such as meal preparation and cleaning, and help support independent living.

Childcare: AI chips also have many possibilities in the field of child-rearing: wearable devices equipped with AI chips can constantly monitor the health of infants and toddlers and notify parents of any abnormalities, enabling them to take early action. Furthermore, AI technology as a whole, including voice recognition and natural language processing, can support communication between parents and children. It can also analyze children's learning progress and comprehension levels to propose optimal learning plans. This enables more effective education tailored to individual needs.

1.8 Contents and Structure of This Book

The content and structure described in this book are as follows. We begin with the fundamentals. Chapter 2 describes the components of AI LSI. Next, in Chapter 3, the basic electronic circuit blocks and their realization as digital or analog circuits are explained. In Chapter 4, based on the elements in Chapters 2 and 3, the basic configuration of NNs and their realization will be explained. It will show the basic LSI circuitry of NN expressivity and inference functions, the configuration of NN learning function LSIization, and the structure of NNs. Chapter 5 will focus on key technologies, including cloud computing, while focusing on edge AI. Specifically, integrated circuits, low-precision NNs, sparse NNs, hardware optimization, FPGAs, spiking NNs and neuro-morphic computing, embedding learning functions, and quantum computing in edge AI.

The above provides a general overview of AI chips, explaining the basic structure of AI LSIs, electronic circuit elements, NN structures, and important technologies for edge AI and cloud computing, so that readers can gain a basic understanding of AI chips.

In the following Chapters 6 and 7, we focus on an Ising machine that solves combinatorial optimization problems as a concrete example of a circuit configuration similar to that of an AI chip. The processing performed by this Ising machine is the same as that of an AI chip for inference processing. Chapter 6 begins with an overview of Ising models and describes the various Ising machine developments underway. Specific topics include an explanation of the Ising model, the relationship between ferromagnetic materials and the Ising model, applications of the Ising model to combinatorial optimization problems, quantum Ising machines, Ising machines using CMOS integrated circuits, and components of Ising LSIs. In Chapter 7, the fully coupled

Ising machine LSI developed in the author's laboratory is presented as a case study. The specific contents of the chapter include the configuration of the fully coupled Ising LSI, the configuration of the large-scale Ising machine LSI, and the scalable configuration.

This is followed by a discussion of memory technologies and their application methods that are essential for improving the performance of these AI chips. Chapter 8 begins the description with the hierarchical structure of memory in computing, followed by an overview of major memory technologies in Chapter 9. Specifically, the hierarchical structure of memory in signal processing and major semiconductor memories (SRAM, DRAM, flash memory, MRAM, FeRAM, PCRAM, and RRAM) is described. Chapter 10 addresses near-memory computing and in-memory computing, which are the foundation for utilizing memory technology in AI chips at the edge.

The final chapter, Chapter 11, will show the evolution and future development of these techniques. Future prospects, including advances in brain-based information processing and integration with other technologies, will also be discussed.

In this book, we will learn how to implement the batch processing of a large amount of data necessary for AI processing in semiconductors, mainly in the wired logic control method, which is a method that uses physical wiring to configure logic circuits and perform operations. This is a matrix calculation (mainly matrix product), and since the data content is placed in a memory element and its processing is integrated with it, it can be understood that this is, so to speak, spatial programming in terms of the element arrangement of the LSI. It is also important that the LSI is scalable. In addition, the key is to take advantage of the fact that the computational processing is not uniform when viewed spatially and to make good use of memory circuit elements. The reason why the utilization of memory circuit elements is important is that the transfer of data between the part that performs the calculation processing and the part that stores the data determines the performance and power consumption of the processing in the LSI. This book also covers basic topics such as how to realize digital and analog circuits, and the connection to field reconfigurable LSIs called Field Programmable Gate Array (FPGA) and neuromorphic chips, which aim to process information in a brain-like manner.

1.9 Summary

AI is a technology that aims to mimic human thought processes or that is perceived by humans as intellectual, and in recent years, has become integrated with machine learning using NNs, especially deep learning. The AI chip is a specialized integrated circuit designed to perform these computational processes in a fast and efficient manner. In order to execute operations (matrix product, etc.) in NNs at high speed, AI chips are equipped with a large number of arithmetic units and are configured to take advantage of parallel processing. In this book, we will learn mainly about the hardware-level configuration, especially the wired logic control method, which is a method that uses physical wiring to configure logical circuits and perform operations. AI chips employ a two-dimensional array-type architecture specialized for matrix operations, which allows them to efficiently process large amounts of data. By enabling advanced data processing not only in the cloud but also at the edge, such AI chips are expected to realize energy-saving and real-time processing, thereby contributing to the construction of a sustainable society.

References

1. John McCarthy, Marvin L. Minsky, Nathaniel Rochester, and Claude E. Shannon. 2006. A Proposal for the Dartmouth Summer Research Project on Artificial Intelligence: August 31, 1955. *AI Magazine*, Vol. 27, No. 4, pp. 12–14. DOI: 10.1609/aimag.v27i4.1904
2. M. Minsky, "Steps toward Artificial Intelligence," *Proceedings of the IRE*, vol. 49, no. 1, 1961. https://doi.org/10.1109/JRPROC.1961.287775
3. E. A. Feigenbaum, "The Art of Artificial Intelligence: Themes and Case Studies of Knowledge Engineering," *Proceedings of the 1980 Conference on Artificial Intelligence*, 1980, pp. 101–109. https://dl.acm.org/doi/10.5555/1622943.1623042
4. S. Haykin, *Neural Networks: A Comprehensive Foundation*, 2nd ed. Upper Saddle River, NJ, USA: Prentice Hall, 1998.
5. F. Rosenblatt, "The Perceptron: A Probabilistic Model for Information Storage and Organization in the Brain," *Psychological Review*, vol. 65, no. 6. https://psycnet.apa.org/doi/10.1037/h0042519
6. K. Fukushima, "Cognitron: A Self-Organizing Multilayered Neural Network," *Biological Cybernetics*, vol. 20, no. 3–4, pp. 121–136, 1975. https://doi.org/10.1007/BF00342633
7. J. J. Hopfield, "Neural Networks and Physical Systems with Emergent Collective Computational Abilities," *Proceedings of the National Academy of Sciences of the United States of America*, vol. 79, pp. 2554–2558, 1982. https://doi.org/10.1073/pnas.79.8.2554
8. J. J. Hopfield and D. W. Tank. ""Neural" Computation of Decisions in Optimization Problems," *Biological Cybernetics*, vol. 52, pp. 141–152, 1985. https://doi.org/10.1007/bf00339943
9. D. E. Rumelhart, G. E. Hinton and R. J. Williams, "Learning Representations by Back-Propagating Errors," *Nature*, vol. 323, no. 6088, pp. 533–536, 1986. https://www.nature.com/articles/323533a0
10. A. Krizhevsky, I. Sutskever and G. E. Hinton, "ImageNet Classification with Deep Convolutional Neural Networks," Advances in Neural Information Processing Systems (NIPS), vol. 25, pp. 1097–1105, 2012. https://doi.org/10.1145/3065386

11. Y. LeCun, Y. Bengio and G. Haffner, "Gradient-Based Learning Applied to Document Recognition," *Proceedings of the IEEE*, vol. 86, no. 11, pp. 2278. https://doi.org/10.1109/5.726791
12. H. Xiao, K. Rasul and R. Vollgraf, "Fashion-Mnist: a Novel Image Dataset for Benchmarking Machine Learning Algorithms," arXiv preprint, arXiv:1708.07747, 2017.
13. J. L. Hennessy and D. A. Patterson, *Computer Architecture: A Quantitative Approach*, 6th ed. Morgan Kaufmann, 2022. Burlington, Massachusetts.
14. M. A. Islam, M. Y. Arafath and M. J. Hasan, "Design of DDR4 SDRAM controller," 8th International Conference on Electrical and Computer Engineering, Dhaka, Bangladesh, 2014, pp. 148–151, doi: 10.1109/ICECE.2014.7026950.
15. H. Jun et al., "HBM (High Bandwidth Memory) DRAM Technology and Architecture," 2017 IEEE International Memory Workshop (IMW), Monterey, CA, USA, 2017, pp. 1–4, doi: 10.1109/IMW.2017.7939084.
16. Tae-Young Oh, Hoeju Chung, Young-Chul Cho, Jang-Woo Ryu, Kiwon Lee, Changyoung Lee, "A 3.2Gb/s/pin 8Gb 1.0V LPDDR4 SDRAM with integrated ECC engine for sub-1V DRAM core operation," 2014 IEEE ISSCC, San Francisco, CA, USA, 2014, pp. 430–431, doi: 10.1109/ISSCC.2014.6757500.
17. Chulhoon Kim, Chanho Lee, "Design of eMMC controller with multiple channels," 2016 International SoC Design Conference (ISOCC), Jeju, Korea (South), 2016, pp. 317–318, doi: 10.1109/ISOCC.2016.7799805.

2

Overview of Artificial Intelligence Hardware LSI (AI Chips) and Its Components

2.1 Information Processing in the Brain

As mentioned in Chapter 1, AI processing is "information processing that humans perceive as intellectual". Subsequently, with the development of neural networks, which began with the study of information processing in the brain, AI processing has come to be treated almost synonymously with processing using neural networks [1] since the advent of deep learning. Neural network technology has greatly advanced pattern recognition, combinatorial optimization, and control. Here is an overview of the brain's information processing and the neural networks based on this.

The brain, which is capable of higher-order thinking, perception, and learning, processes information through neurons, which are nerve cells, and their connections. Neurons transmit signals to each other through connections called synapses, forming a network in the brain. This network allows information to be processed in parallel and enables the mechanisms of pattern recognition and learning. The brain's remarkable capacity to receive input information, interpret it, and generate appropriate responses and actions is enabled by the function of these neurons and the neural networks they form.

Let us examine this in some detail. Figure 2.1 shows this neuron and its connections schematically [2]. The brain is composed of a special group of cells called neurons, which are the basic unit of information processing. Neurons consist of three main parts. The cell body (Soma) is the central part of the neuron, contains the cell nucleus, and is responsible for maintaining cellular metabolism and function. Dendrites are numerous projections from the cell body that receive signals from other neurons. On the surface of dendrites, contact points, called synapses, are formed where information is passed on and off. Axons are long projections from the cell body that transmit electrical signals (action potentials) to other neurons and muscle cells. Axons terminate in presynaptic terminals, where neurotransmitters are released and information is transmitted. The brain's sophisticated information processing is achieved by changes in the connections between neurons at synapses.

In this process, electrical signals are converted into chemical signals at the synapse and transmitted to the next neuron. This process proceeds as follows. First, an action potential is generated, which reaches the presynaptic

 DOI: 10.1201/9781003599067-2

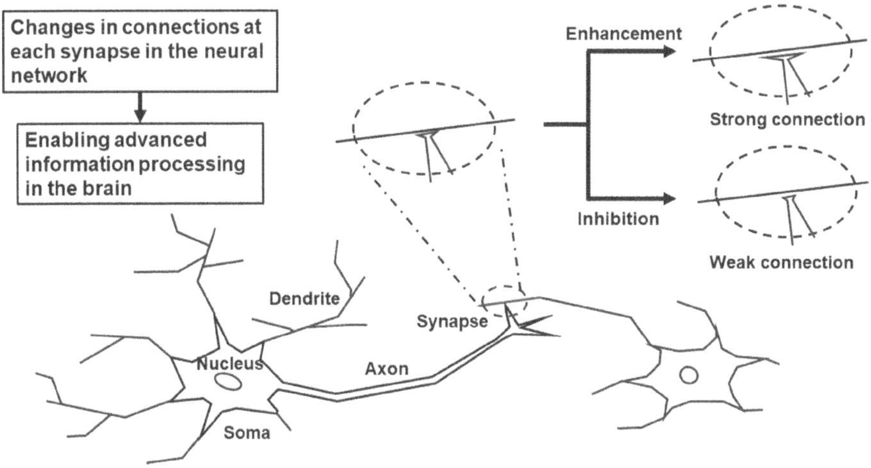

FIGURE 2.1
Brain's information processing unit.

terminal through the axon when the neuron is excited. Next, the action potential opens a calcium channel in the presynaptic terminal, allowing calcium ions to flow in. This causes synaptic vesicles to fuse with the cell membrane, and neurotransmitters are released into the synaptic cleft. This released neurotransmitter binds to receptors on the dendrites of the postsynaptic cell and either enhances or inhibits the next neuron. In excitation, when one neuron sends a signal to another neuron, it promotes the excitation of the other neuron. This strengthens the synaptic connections and facilitates the transmission of stronger signals. Excitation plays an important role in learning and memory formation. In inhibition, on the other hand, when a neuron sends a signal to another neuron, it inhibits the excitation of the other neuron. Inhibited connections make it more difficult for information to be transmitted and serve to prevent overexcitation of the neural circuit.

These potentiation and inhibition mechanisms result in strong and weak connections between neurons. And it is believed that the brain performs sophisticated information processing by means of these changes in connections. Although the actual functions of the brain have not been fully elucidated, it has become clear that artificial neurons and artificial neural networks, as discussed in the next section, which were devised based on changes in these neurons and their connections, can produce appropriate output responses to input by adjusting the strength (weight) of the connections. Additionally, it is possible to obtain output with added value based on input information [1], rather than a simple relationship between input and output. In other words, it has become clear that useful information processing can be achieved. Furthermore, at present (2025), although it is unlikely to be the actual information processing of the brain, the possibility of technology that can mimic and even surpass this is being discussed as artificial general intelligence (AGI).

2.2 Components of an AI Chip

Based on the findings of the previous section, a neural network LSI (large-scale integrated circuit) is constructed. To summarize the components of this in advance, there are only the following two:

1. Computational units with special functions called artificial neurons
2. Networks constructed by connecting multiple artificial neurons (very large number of connections required)

Based on this, the neural network represented by modern deep learning (deep learning: machine learning with multilayer neural networks) is constructed. Based on these simple arithmetic elements, a method for neural networks to learn from data (generally backpropagation) has been developed to enable data processing and pattern recognition. By using LSI technology here, the computational power of this neural network can be increased, and efficient data processing can be achieved.

These functions can, of course, be written in software and realized using personal computers (PCs) and the cloud. However, the volume of data continues to expand, and advances in hardware with superior power-to-performance ratios have become critical. In addition, the development of IoT technology, which is also a major factor in this data volume expansion, requires immediate on-the-spot processing of those data, i.e., processing at the edge. In other words, artificial intelligence processing at the edge is important. Here, the development of hardware that can be implemented at low power and low cost is expected. Hardware with performance (high speed, low power, low cost) that exceeds conventional computing (central processing unit+memory) is needed.

In this context, there is generally a trade-off between implementation area (cost), power consumption, and speed. In the technology covered in this book, the goal is to find a fast, low-power, and low-cost solution that breaks this trade-off by using integrated circuit chips specialized for AI processing. In particular, there is a growing demand for integrated circuit chips that specialize in the functions required for deep learning. These chips are equipped with a large number of arithmetic units and are capable of processing many operations in parallel at the same time. Furthermore, the specialized circuit design can achieve high computational processing power and power efficiency.

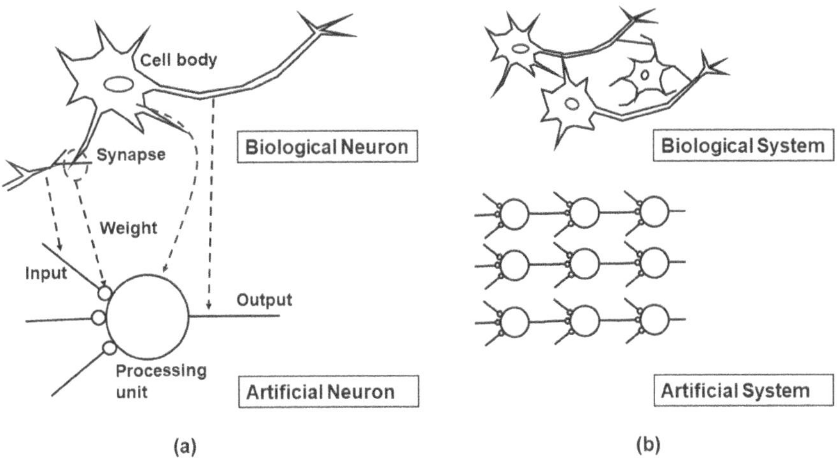

FIGURE 2.2
Biological and artificial neurons, and networks. (a) Neurons and (b) neural networks.

2.3 Artificial Neurons

To repeat, a neural network consists of a number of basic arithmetic units with a certain function called artificial neurons, which are connected together. Here, the artificial neuron has the configuration shown in Figure 2.2a, receives input signals, processes them, and generates output signals. This then constitutes a neural network, as shown in Figure 2.2b. The advantage of the neural network is that it is a network with only this basic configuration. This is a well-known configuration, but from the aspect of LSI circuits, it is possible to realize various functions as artificial intelligence by arranging and connecting elements of the same type, which is extremely important because it is easy to design. In terms of operation, this also means that many parts can be processed in parallel. This means that the power-to-performance ratio can be improved by using LSIs. Furthermore, although this may lead to an aggressive reduction in arithmetic precision (which will also lead to lower power and area), the neural network processing may not require all elements to be fully operational.

This neural network can also be used for recognizing numbers or, with the change of parameters (weights and biases), it can be used to classify images. It becomes, so to speak, a reconfigurable integrated circuit. Furthermore, if the relationship between input and output is called a function, the neural network has great expressive power as this function. In other words, it appears to be able to represent the relationship between any input and output. On the other hand, this also means that if you decide on a target and specialize in it, redundant parts will appear inside. In other words, some operations that do not contribute much to the output result will appear. Taking advantage of this property of neural networks, LSI circuits are required to minimize power consumption, maximize speed, and minimize implementation area.

Figure 2.3a is an example of a description of a model that focuses on the computation [2] performed there by an artificial neuron. One of these artificial neurons, again in the language of integrated circuits, is a device that performs the following functions. The function 2) enables the neural network to achieve its great expressive power as a function.

1. Sum-of-products (weighted sum) operation performed on data with multiple inputs and weights
2. Calculation of activation functions to realize nonlinear transformations performed on the above results

As shown more concretely in Figure 2.3b, this artificial neuron i has n inputs $x_j(j=1\sim n)$ and one output y_i. Also, the input x_j is not directly connected to this artificial neuron i directly, but is weighted. This weighting is denoted as $w_{ij}(j=1\sim n)$. The artificial neuron takes the product of these two values and then sums the result. If the input signal is an analog signal, the product of the two is that the analog input signal multiplied by the resistance that is the weighting, and input x_j is transmitted, but if the resistance is high, the coupling is weak, meaning that the weighting for this input is small and attenuated. For digital signals, a low resistance corresponds to a stronger w_{ij} value for the input signal, while a high resistance corresponds to a weaker

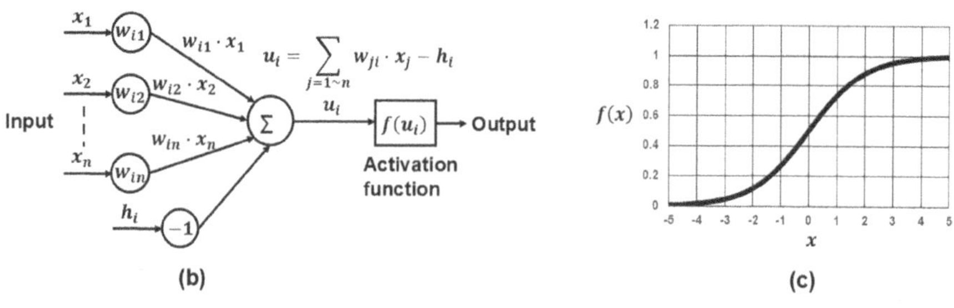

FIGURE 2.3
Processing in artificial neurons. (a) Processing unit and (b) activation function.

w_{ij} value. The sum of these inputs is taken. h_i represents the bias or threshold value and is responsible for adjusting the output.

The above is shown as a sequential operation from input to output. First, the sum-of-products operation of w_{ji} and x_j is performed. Setting this result as u_i is the following equation.

$$u_i = \sum_{j=1}^{n} w_{ij} \cdot x_j - h_i \tag{2.1}$$

This value determines to what extent the result of the sum-of-products operation of w_{ij} and x_j is considered as a whole. Here, the $-h_i$ term in equation (2.1) may be $u_i = \sum_{j=0}^{n} w_{ij} \cdot x_j$ with the input $x_0 = h_i$ and the weights $w_0 = -1$. The fact that this can be summarized by equation (2.1) forms the basis for learning through the error backpropagation described later. In other words, equation (2.1) is linear with respect to the input x_j and the input after weighting with the weights.

After this, through the activation function f, which is also shown in the next section, the output y_i is obtained from equation (2.2).

$$y_i = f(u_i) \tag{2.2}$$

The activation function $f(x)$ has an output that is nonlinear with respect to the input x, as shown in Figure 2.3c. The range of this output is small with respect to the input. Also, and this is a recurring note, the nonlinear process gives the neural network greater expressive power as a function. This allows it to learn complex patterns. Also, in fact, the activation function renders many neurons inactive. However, functions such as recognition are possible in the neural network as a whole. This is important when improving power efficiency in circuit operation.

2.4 Neural Networks

Next, this neuron is used to form a network. Multiple neurons form a layer, and based on this, a multilayered structure is called a neural network. This layer structure consists of three main types of layers: input layer, hidden layers, and output layer. At this time, by having multiple hidden layers, neural networks can achieve more complex functions. Neurons in the hidden layer perform calculations based on information from the previous layer and transmit the information to the next layer. The more of these layers there are, the "deeper" the network is, and this is the basis of the technology known as deep learning.

As shown in Figure 2.4a, there are m neurons i for multiple but identical pairs of inputs. That is, $u_i(i = 1 \sim m)$. In a neural network, this m pair of neurons is treated as a single layer. For each of m neurons, $u_i = \sum_{j=0}^{n} w_{ij} \cdot x_j$.

Here, we note x_j and u_i as vectors and w_{ij} as matrices in the form of matrix equations, as it will be useful later when considering the spatial arrangement of the electronic circuit. In other words, for $x_j(j = 1 \sim n)$ and $u_i(i = 1 \sim m)$ and $w_{ij}(i = 1 \sim m, j = 1 \sim n)$, we have the following.

$$\begin{pmatrix} u_1 \\ u_2 \\ u_3 \\ \vdots \\ u_m \end{pmatrix} = \begin{pmatrix} w_{11} & w_{12} & w_{13} & \cdots & w_{1n} \\ w_{21} & w_{22} & w_{23} & \cdots & w_{2n} \\ w_{31} & w_{32} & w_{33} & \cdots & w_{3n} \\ \vdots & \vdots & \vdots & \ddots & \vdots \\ w_{m1} & w_{m2} & w_{m3} & \cdots & w_{mn} \end{pmatrix} \begin{pmatrix} x_1 \\ x_2 \\ x_3 \\ \vdots \\ x_n \end{pmatrix} \tag{2.3}$$

In the in-memory computing of Chapter 10, this matrix arrangement of w_{ij} also corresponds to the spatial arrangement of the memory that stores the weights w_{ij}, which is a matter of ingenuity for memory circuit designers.

Using this as a unit, a neural network has multiple layers, as shown in the example in Figure 2.4b. The input layer is the first layer of the neural network and receives data from the outside. The hidden layer is located between the input and output layers and is responsible for extracting features of the data. By having multiple hidden layers, more complex models can be built. The output layer is the final layer of the neural network and outputs the processing results. In classification problems, it is common to output the probability for each class.

The input layer as a circuit is the function of holding input information and conveying it to the next layer, so it requires a register circuit and an

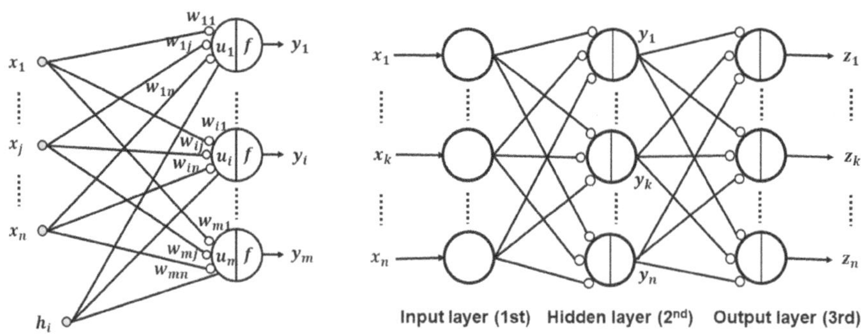

(a) (b)

FIGURE 2.4
Construction of neural network. (a) Connections with multiple neurons in one year and (b) multilayering (when each layer has the same number of neurons).

output driver circuit that can drive the inputs of the artificial neurons in the next layer. Since the hidden layer consists of hundreds or more artificial neurons, the output driver will be connected to these inputs. From the viewpoint of integrated circuits, it is important to devise the circuit configuration in such connections of neural networks. For example, in the case of analog-like weights w_{ij}, the advantage of handling with analog circuits can be utilized. On the other hand, to construct a large-scale network, it may be advantageous to use digital circuits for ease of construction and robustness of circuit operation.

Next, let us look at the sum-of-products and nonlinear transformation operations one by one.

2.5 Multiply-and-Accumulate Operation

In neural network processing, the above equation (2.1) is its basis, and here, looking at the $u_i = \sum_{j=1}^{n} w_{ij} \cdot x_j$ part, as a circuit, it is a mass of a sum-of-products arithmetic circuit. As a digital signal, it is an information processing operation to find the product of this with respect to the input x_j, but as an analog signal, a resistor to the input x_j will be inserted.

When realized with the digital circuit shown in Figure 2.5a, this sum-of-products processing has the advantage of high accuracy and repeatability. However, although digital circuits are suitable for high-speed calculations, they have the disadvantage of increasing power consumption as the amount of data

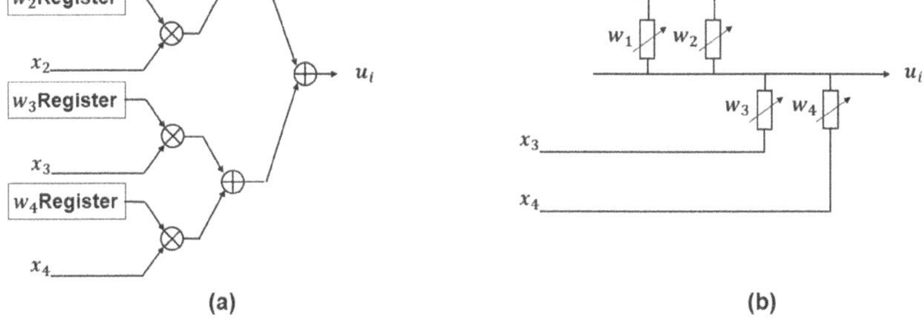

FIGURE 2.5
Examples of computations in artificial neurons. (a) Digital and (b) analog.

to be processed increases. In addition, the large number of logic gates required for the sum-of-products operation increases the circuit mounting area. On the other hand, when the analog circuit shown in Figure 2.5b is used to realize the sum-of-products operation, it is said to have the advantages of high-speed computation and superior energy efficiency compared to digital circuits, depending on how the circuit is devised for a specific purpose. Furthermore, since fewer arithmetic elements are required for the sum-of-products operation, the circuit mounting area can be reduced and the level of integration can be expected to improve. However, analog circuits are not suitable when high accuracy is required, since errors occur in the addition process.

If a large number of these circuits are combined, a neural network can be constructed. This means that a neural network can be constructed by using a regular circuit array of basic circuits, which leads to good compatibility with in-memory computing. In-memory computing is a technology that achieves high-speed computation by integrating memory and computation in the same circuit. However, it requires expertise in circuit design because it deals with memory, and it also requires advanced manufacturing technology for integration with memory elements.

2.6 Nonlinear Transformation Operations (Activation Functions)

Next, the artificial neuron performs the nonlinear transformation operation (activation function), Equation (2.2), on the result of the sum-of-products operation. In other words, an operation using the activation function is performed on u_i, the result of the sum-of-products operation of w_{ij} and x_j. The result is the output y_i, which is the input to the next artificial neuron. The output or input of the previous neuron is weighted, and a nonlinear transformation is added to the summed result. This operation allows the neural network to exhibit complex functional relationships for inputs and outputs. Without the activation function, the neural network would be merely a linear transformation and would not be able to capture complex patterns of data. Using a neural network with an activation function, for example, it is possible to realize a function (i.e., a function) that maps the input of handwritten letter patterns from the numbers 0 to 9 to the output of which number the pattern is. It is possible to recognize (match) which number is represented by scanned handwritten numbers.

As for this activation function, if u_i is represented as the variable x (horizontal axis) and y_i as the variable$y$ (vertical axis), the function $y = f(x)$ has the form shown in the characteristic diagram in Table 2.1. As shown in this figure, the activation function also serves to limit the output of the neuron. In this table, the derivatives $f'(x)$ are also shown, which are necessary when the neural network is trained.

Let us look at these activation functions one by one, albeit briefly.

TABLE 2.1
Activation Function Example

Function Name	Diagram	$f(x)$	$f'(x)$
Step	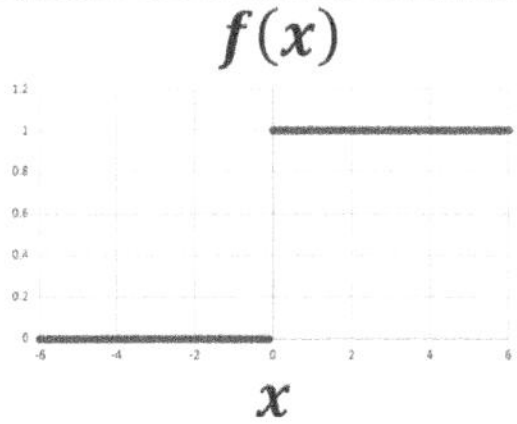	$\begin{cases} 0, & x < 0 \\ 1, & x \geq 0 \end{cases}$	$\delta(x)$
Sigmoid	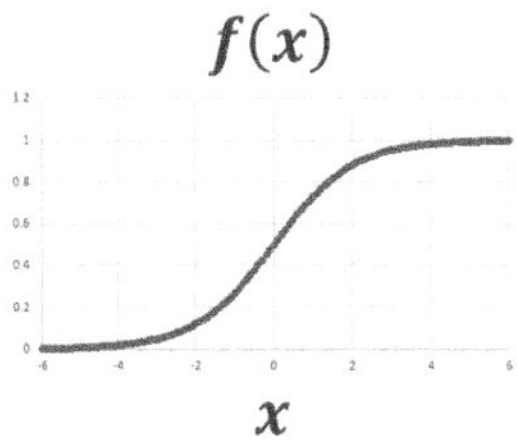	$\frac{1}{1+e^{-x}}$	$f(x)(1-f(x))$
Tanh	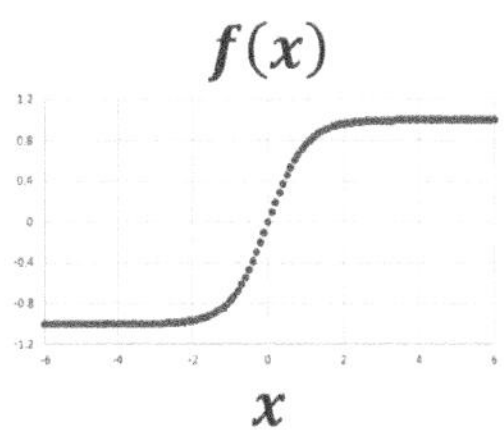	$\frac{e^{x}-e^{-x}}{e^{x}+e^{-x}}$	$1-f(x)^2$
ReLU	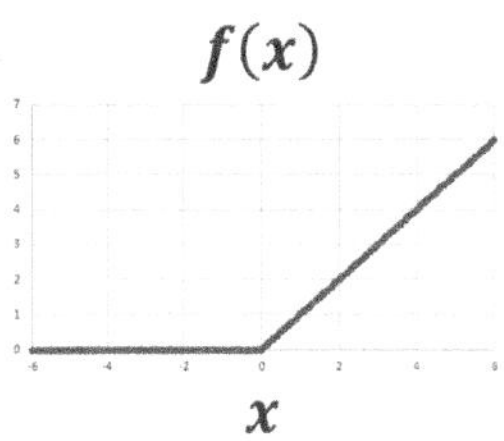	$\max(0,x)$	$\begin{cases} 0, & x < 0 \\ 1, & x \geq 0 \end{cases}$
Softmax	–	$\frac{e^{x_i}}{\sum_{k=1}^{n} e^{x_k}} (= (y_i))$ $(i=1,2,\cdots,n)$	$\frac{\partial y_i}{\partial x_j} = y_i(\delta_{ij} - y_i)$

The step function is a very simple function that outputs 1 if the input exceeds a certain threshold and 0 if it does not. In integrated circuits, this function is easy to realize and can be performed at low cost and high speed. However, because the step function is non-differentiable, it is not suitable for some learning algorithms, such as error backpropagation.

The sigmoid function is a function that converts input values from 0 to 1. The function is differentiable and suitable for learning algorithms, such as error back propagation, because its output varies smoothly with the input value. The challenges are that the sigmoid function contains an exponential function, which makes it difficult to execute at high speed unless it is devised in integrated circuits, and that it suffers from gradient vanishing, where the gradient becomes almost zero as the input value increases.

The Tanh function is similar to the sigmoid function, but has an output range of –1 to 1. Tanh functions have larger gradients than sigmoid functions, so there is less of a gradient loss problem. However, like the sigmoid function, it contains an exponential function, which makes it difficult to execute at high speed without some ingenuity in integrated circuits.

The ReLU function is an activation function that outputs 0 when the input is negative and outputs the input as is when the input is positive and is currently the most widely used. In integrated circuits, it can be implemented as a simple comparison operation and is capable of high-speed operations.

The softmax function is used slightly differently than other functions, but it is used to normalize input values to produce a probability distribution. It is used in the output layer of a classification problem. By setting the range of input values to [0,1] and adjusting them so that they sum to 1, it is possible to represent the probability of each class in a classification problem. However, the integrated circuit implementation requires complex arithmetic operations and ingenuity.

Hardware implementation of the activation function will increase computation speed and improve energy efficiency. It enables real-time data processing, which is essential, especially in edge devices and embedded systems, where quick response is required. In implementation, however, the design can be complex, especially for nonlinear functions. Analog circuits are suitable for realizing functions such as the sigmoid function, but designing analog circuits requires skill. Here, the sigmoid function itself may be simplified and devised in a digital circuit; the ReLU function is relatively easy to implement in hardware and saves computational resources.

2.7 Chips for Learning and Inference

In this way, the artificial intelligence LSI (AI chip) described in this book is an integrated circuit that realizes an artificial neural network, consisting of a large number of artificial neurons (nodes) and the connections (weights) that connect them.

Here, there were two processes in neural networks: learning and inference. Based on a large amount of data, the network learns by combining the weights of the nodes and connections of the neural network. On the other hand, inference is the process, in which the network learns the characteristics of the data in learning and applies them to the data it wants to identify or classify. There are also two types of AI chips that handle this process, one for learning and the other for inference.

The learning chip requires a high level of computational power due to the calculations needed to extract features from large amounts of data. Large memory is also required. Currently, GPUs are widely used. This enables parallel processing and reduces the time required for processing. In addition, the process of extracting features from this data mainly involves matrix operations and tensor calculations, as described later. The current challenge is that processing is currently performed using GPUs, but power consumption is becoming excessive. For this reason, energy-efficient LSI chips are desired for artificial intelligence processing.

Chips for inference will be equipped with a learned neural network to make decisions on new data. They are also expected to be used at the edge. It can process data without relying on the cloud, protect privacy, and reduce communication costs. This is an area where realization in field programmable gate arrays (FPGAs) and application-specific integrated circuits (ASICs), described later, is being sought first. Since these edge devices are often used in resource-limited environments, they are required to have not only computing power but also low power consumption, guaranteed operating temperatures in actual applications, and ease of maintenance. However, there are many applications that require real-time response, such as self-driving cars and recognition systems.

AI chips are expected to bring progress in AI processing by designing circuits specialized for these learning and inference tasks. To this end, they are required to have high computational and parallel processing power, high accuracy, low power consumption, and efficient data transfer to and from memory, as well as flexibility and scalability to support various architectures and models, and support as an edge device.

2.8 Analog and Digital Circuits

We mentioned that there are cases where the signal in a neural network is treated as a digital or analog signal. Although not necessarily consistent with this signal representation, it is important to note whether the realization of the neuron's function is done with analog or digital circuits [3]. Table 2.2 summarizes the comparison. In some cases, the use of analog circuits can reduce the size of the circuit by an order of magnitude or more compared to digital circuits. This is because the nonlinear operations that appear in neural networks can be implemented with a small number of transistors. Power consumption can also be reduced. In addition, in the processing that integrates memory elements and arithmetic operations, which will be discussed in Chapter 10, operation as an analog circuit is the base. This enables a significant improvement in processing performance and lower power consumption, which is accelerating development. In addition, since memory

TABLE 2.2

Comparison of Digital and Analog Circuits in AI Chip Configurations

Circuit Type	Circuit Size	Power Consumption	Low Voltage	Noise Tolerance	Robustness to Device Variations	Ease of Design	CMOS Process Portability
Digital	Large	Small	○	○	○	○	○
Analog	Small (depending on circuit configuration)	Small (depending on circuit configuration)	△	△	△	△	×

technology itself cannot be handled without specialized knowledge and special techniques, this is a significant advantage for AI chips.

On the other hand, lowering the supply voltage is effective in reducing power consumption, but is difficult in normal analog circuit operation. In addition, digital circuits are less susceptible to substrate noise, power supply fluctuations, and process variations. Analog circuits can also minimize these effects, but power consumption and circuit mounting area increase. Furthermore, analog circuits generally require dedicated designs for all circuits, while digital circuits are increasingly automated, such as being generated from logic descriptions. In addition, analog circuits must be redesigned when CMOS process technology changes, but digital circuits can be ported relatively easily.

2.9 Summary

Modern artificial intelligence processing systems are constructed on neural networks inspired by the information processing of the human brain. This network consists of a substantial number of interconnected computational units called artificial neurons. To implement this as an LSI (large-scale integrated circuit), the main components required for artificial neurons, namely sum-product (weighted sum) circuits and activation function circuits that realize nonlinear transformation, are important. The sum-of-products operation adds the results of multiplying multiple inputs and their corresponding weights, and the activation function adds a nonlinear transformation to the output. The activation function can be of the form sigmoid, Tanh, ReLU, and other forms. The processing of neural networks is predominantly categorized into two distinct phases: "learning" and "inference." The LSI for learning requires high computing performance and a large memory. On the other hand, inference is often performed in resource-limited environments, such as at the edge, where low power consumption and high-speed operation are important.

References

1. I. Goodfellow, Y. Bengio and A. Courville, *Deep Learning*. Cambridge: The MIT Press, 2016.
2. E. R. Kandel, J. D. Koester, S. H. Mack and S. A. Siegelbaum., *Principle of Neural Science*, 6th ed. New York: McGraw-Hill Professional, 2021.
3. S. Sedra and K. C. Smith, *Microelectronic Circuits*, 8th ed. Oxford: Oxford University Press, 2019.

3

Basics of Large-Scale Integrated Circuits for AI

The neural network (artificial intelligence) LSI (AI chips) seen in the previous chapter was, to show again, based on the configuration shown in Figure 3.1a and networked with it. In this book, we learn to realize this as an LSI circuit. Of course, it is possible to build a neural network in software and process it with an ordinary PC such as the one shown in Figure 3.1b. However, the power/performance ratio that can be achieved with this configuration based on data transfer between processor and memory is not suitable for advanced AI processing at the edge in the future. On the other hand, an LSI circuit that performs the processing in Figure 3.1a as an AI chip, as described in Chapter 1, has an array structure with multiple processing elements (PEs) laid out in two dimensions, as shown in Figure 3.1c. In this AI chip, relatively large amounts of data are processed at once. Therefore, although memory may be incorporated in each PE or transferred from other PEs, high speed and low power consumption can be achieved by processing in this array structure.

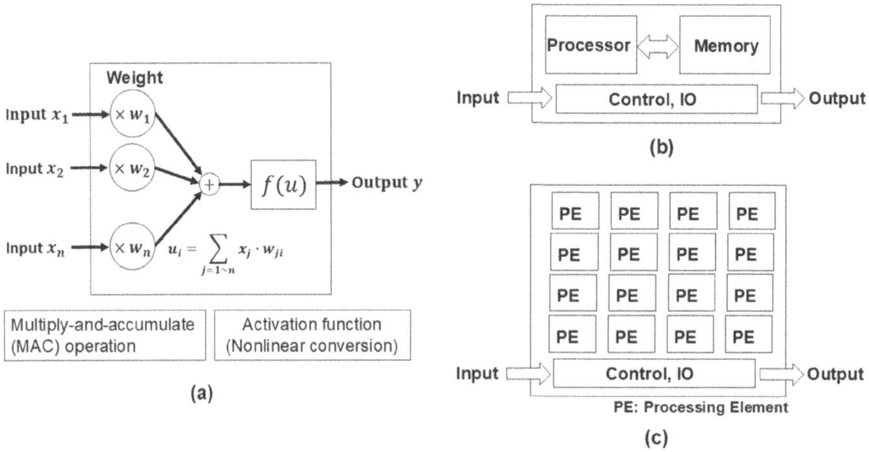

FIGURE 3.1
Functions of neural networks and basic configuration for implementing them in LSI as AI chip. (a) Basic processing of neural networks, (b) processing on conventional CPUs, and (c) processing on AI chips.

DOI: 10.1201/9781003599067-3

In this section, we summarize what kind of LSI circuits realize these PEs. There are two types of circuits: digital circuits and analog circuits, but mainly digital circuits are widely used. However, analog circuits may be able to achieve high power-to-performance ratios for certain purposes. In addition, the method of realization in analog circuits is the original form of circuit operation using memory elements for in-memory computing.

3.1 CMOS LSI Circuit

In LSI circuits [1], CMOS technology is used for its high performance, low power consumption, and ease of miniaturization. The basic gate circuits used to construct LSI digital circuits using CMOS technology are inverter circuits, NOR circuits, and NAND circuits. With these circuits, all the logic circuits necessary for an LSI with the desired functionality can basically be realized.

Figure 3.2 shows a CMOS inverter circuit. Figure 3.2a shows the symbols, truth table, and circuit diagram of the inverter circuit. If the input signal is 1, the output is 0; if the input signal is 0, the output is 1. Figure 3.2b shows its cross-sectional structure. The inverter circuit consists of a p-channel MOS transistor (pMOS) and an n-channel MOS transistor (nMOS). Depending on the input signal, the source and drain of the nMOS are opposite each other, so that when the source (s) and drain (d) of the pMOS are conducting, the source and drain of the nMOS are nonconducting. This complementary structure prevents current flow except when the input is inverted. This allows the inversion of the input signal to be obtained at low power and high speed. On the other hand, a through-current can flow when the input is inverted, and to avoid this, it is important that the time transition of the input is fast. Since the signal is transmitted to the next stage, the output transition by the inverter must also be fast. For this reason, it is important to set the current drive capability of the pMOS and nMOS to match the load capacitance of the output. In addition, since signals are transmitted through metal wiring, the

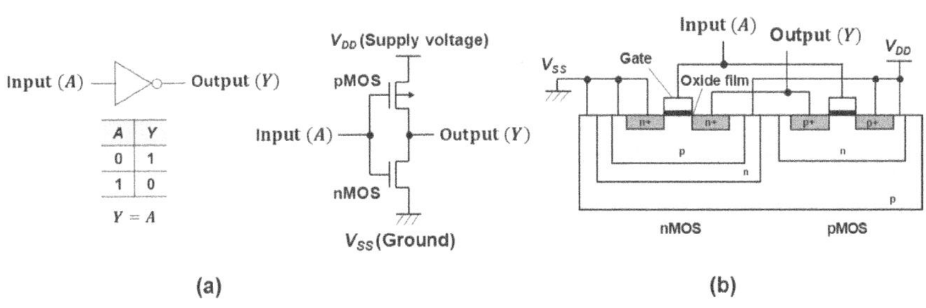

FIGURE 3.2
CMOS inverter. (a) Inverter circuit and (b) cross-section.

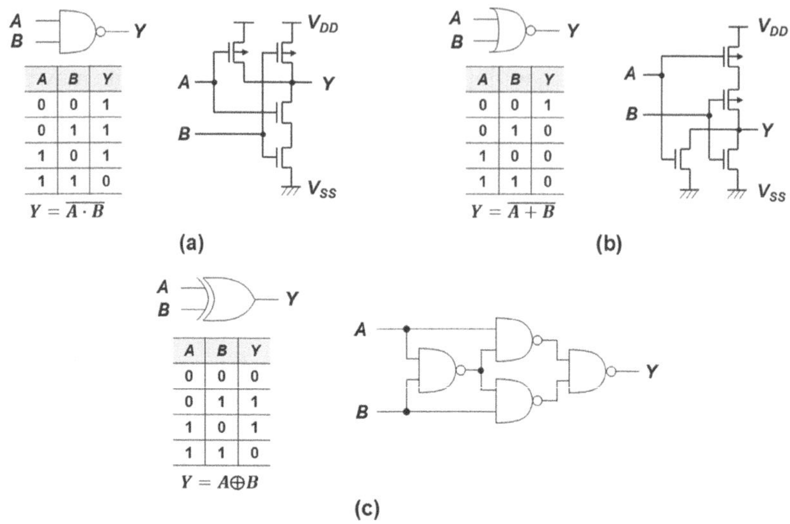

FIGURE 3.3
CMOS logic circuits. (a) NAND circuit, (b) NOR circuit, and (c) EXOR circuit.

time constant (RC) of this wiring must also be small. In general, the rise and fall times of signals are set to values calculated from frequencies 10 times or more than the clock frequency.

Logic circuits can be assembled on this basis. Figure 3.3 shows a typical logic gate circuit, with its symbols, truth table, and circuit diagram for the NAND, NOR, and EXOR circuits as the two-input case. The NAND circuit in Figure 3.3a is a circuit whose output is 1 unless all inputs are 1. Given multiple inputs, this circuit calculates the logical product of those inputs and inverts the result. The NOR circuit in Figure 3.3b is a circuit whose output is 0 unless all inputs are 0. This circuit, when it has multiple inputs, computes the disjunction of those inputs and inverts the result. These two circuits can be used to compute logical products and disjunctions, and various logic functions can be realized by expanding these circuits. The EXOR circuit in Figure 3.3c is a circuit whose output is 1 when either of the two inputs is 1. This circuit is used for data comparison and encoding because it can compute exclusive or. The output is whether the two inputs are different or the same.

3.2 Digital Circuit Block

Among the artificial intelligence integrated circuits based on these components, the main components required to construct a neural network are an adder, a multiplier, a flip-flop/register, a nonlinear transformer, and a random number generator. The arithmetic operations of neural networks and the Ising machines described in Chapters 8 and 9 consist of sum-of-products operations, nonlinear transformations of the results, and stochastic operations using random number generators. The following is a basic 2-bit circuit. In reality, operations are often performed in byte units or more, but extensions should be easy. The circuit examples are based on NAND and NOR circuits. Circuit examples using transmission gates and dynamic operation are omitted.

3.2.1 Adder

An adder is a circuit that takes a binary number with multiple bits and outputs the sum of those bits. When realized in a digital circuit, an adder is realized by a combination of a full adder and a half adder. The half adder shown in Figure 3.4a is a circuit that inputs two 1-bit binary numbers and outputs their sum and carry forward. it has four types of inputs and outputs: 0 + 0 = 0, 0 + 1 = 1, 1 + 0 = 1, 1 + 1 = 0, and 1 round up. Based on this, the full adder depicted in Figure 3.4b is a circuit that inputs two 1-bit binary numbers and the carry from the preceding digit, and outputs the sum and the carry. It can be constructed with two half-adders and an OR circuit.

By combining these 1-bit adders in series, a multibit adder can be constructed. Figure 3.5a is an example of a ripple carry adder, which is a method of sequentially sending carry information to the next bit calculation. Although pipelining can be devised, the computation time increases in proportion to the increase in the number of digits. On the other hand, Figure 3.5b shows a carry-lookahead adder, which generates carry signals using a separate logic circuit to calculate all bits simultaneously, thereby achieving higher speed.

3.2.2 Multiplier

A multiplier is a circuit that outputs the result of multiplying two input values. It is realized by combining multiple adders and registers, as shown in Figure 3.6a. It calculates the product of the parts divided by the AND circuit

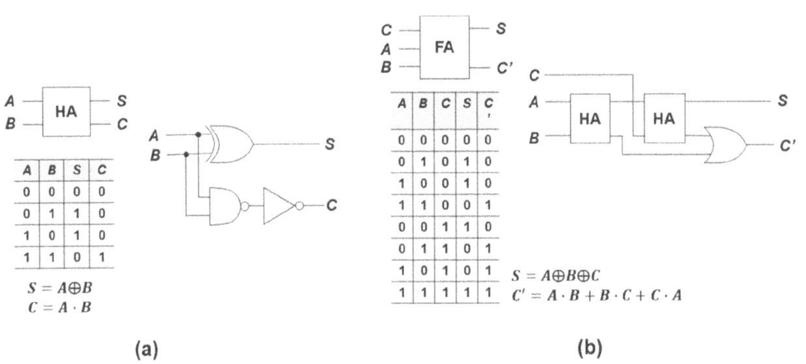

FIGURE 3.4
Adder. (a) Half adder and (b) full adder.

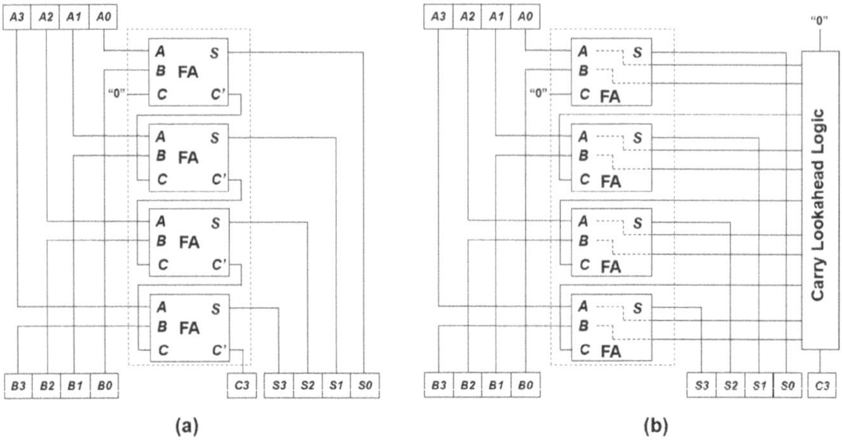

FIGURE 3.5
Multibit adder. (a) Ripple-Carry adder and (b) Carry Look-Ahead adder.

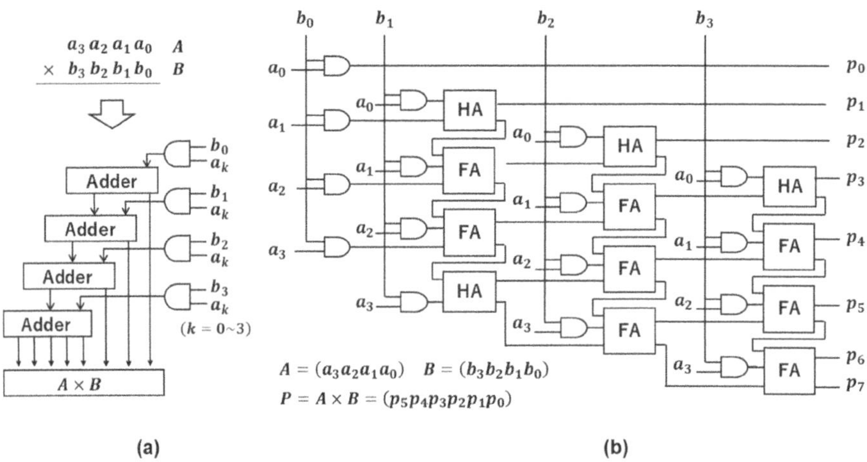

FIGURE 3.6
Multiplier. (a) Multiplier and (b) multiplier circuit configuration.

and adds them. Specifically, when multiplying the multiplicand (A) by the multiplier (B), we first calculate the partial products as follows: for each digit of the multiplier (b_0, b_1, b_2, b_3), if the corresponding digit of the multiplicand (a_0, a_1, a_2, a_3) is 1, we leave it as is. If the multiplicand is 0, we set it to 0. We then shift these values sequentially and arrange them in order, finally adding all the values together. This process is directly implemented in the parallel multiplication circuit shown in Figure 3.6b. For multiplication, a partial product of each digit is obtained by AND, and a parallel adder FA is shifted one bit to the left to add these products. Although this circuit is fast, it requires all adders for the number of product additions, and the larger the number of multiplication bits, the larger the circuit scale. To reduce the number of additions, there are also multiplier circuits that use AND to obtain the partial product of each digit and then shift all adders in parallel to the left one bit to add these products.

3.2.3 Latch Circuit

A latch circuit is a basic circuit used in digital circuits to hold data. A latch circuit can store one bit of information. It accepts an input signal on the rising edge of a clock signal and holds the output signal. The RS latch shown in Figure 3.7a has two inputs (SET and RESET) and one output. When a "Set" signal is given to one input, the output goes High, and when a "Reset" signal is given to the other input, the output goes Low. As shown in the logic circuit, this uses feedback to maintain the state. That is, it is composed of two NAND gates, and the state is held by the output returning to the input. On the other hand, the D latch shown in Figure 3.7b has a single data input and holds the D input signal at its output when the clock signal is on a rising edge.

3.2.4 Register and Shift Register Circuits

A register circuit is realized by using a latch circuit for the required number of digits. For example, an n-bit register can be created using n D latches. It is used to temporarily store data. Additionally, a shift register circuit can perform data input and output in serial or parallel mode, with the capacity to shift data bit by bit.

Figure 3.8a shows a structure in which multiple D latches are connected in series, and each time a common clock signal is input to each D latch, the data stored in one D latch is shifted to the next D latch. The serial input (D) is applied to the data (D) input of the first flip-flop. The data from DI are latched on the rising edge of the clock (CLK) and displayed at Q0. Four clock pulses transfer the data from DI to the fourth flip-flop. As a result, the serial input (DI) is converted to parallel output data appearing at Q0, Q1, Q2, and Q3.

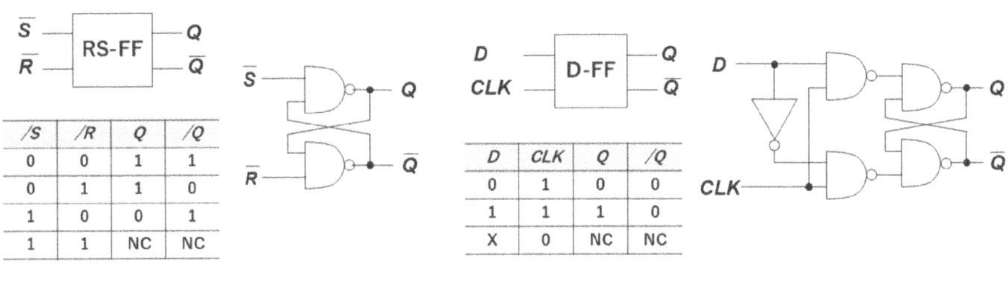

/S	/R	Q	/Q
0	0	1	1
0	1	1	0
1	0	0	1
1	1	NC	NC

D	CLK	Q	/Q
0	1	0	0
1	1	1	0
X	0	NC	NC

FIGURE 3.7
Latch circuits. (a) RS latch (flip-flop) and (b) D latch (flip-flop).

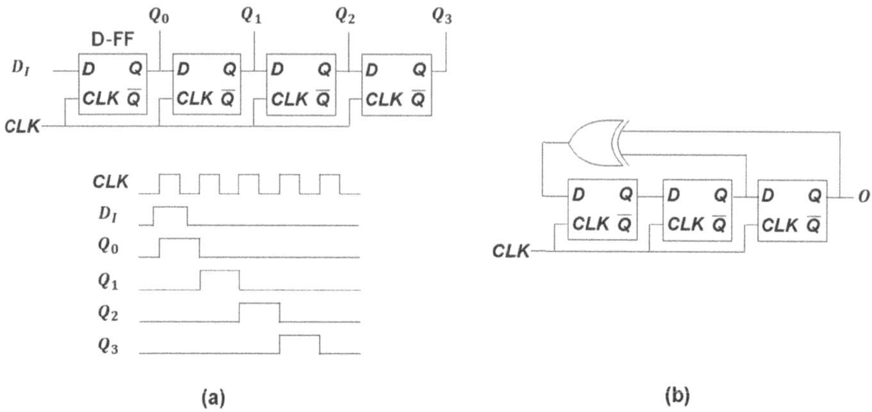

FIGURE 3.8
Register circuits. (a) Shift register and (b) linear feedback shift register (pseudo-random number generator).

3.2.5 Random Number Generation Circuit

Shift register circuits are used in delay circuits, parallel-to-serial conversion circuits, and serial-to-parallel conversion circuits, but they are also used to generate the random numbers needed to perform probabilistic operations that are important in artificial intelligence processing. In order to realize probabilistic operations, it is necessary to generate a random number and compare the magnitude of this value with the value obtained. The generation of this random number is achieved through the implementation of a linear feedback shift register (LFSR), a pseudo-random number generator, as illustrated in Figure 3.8b. LFSR is a shift register that feeds back the register output or the XOR of the outputs to the intermediate stages. The n-bit LFSR register transitions through 2^n-1 states and back. It can generate pseudo-random numbers, called M-sequences. When generated sequentially, there is no fixed pattern in the numbers that are output, but the order of the numbers is fixed.

3.2.6 Nonlinear Conversion Circuit

It is difficult and inefficient to realize a nonlinear conversion circuit in a digital circuit. Therefore, a method is used that refers to a lookup table (LUT) [2] prepared as shown in Figure 3.9a. A LUT is a table that stores output values corresponding to input values in advance. Piecewise linear approximation (PWL) [3], shown in Figure 3.9b, is a method of dividing the activation function into multiple segments and connecting each segment as linear. Although a linear function may be assembled as it is, when a sigmoid function is realized by LUT, the input value may be divided by a step width corresponding to the bit width, and the output value of the sigmoid function at that division point may be stored in the LUT. The activation function can then be realized

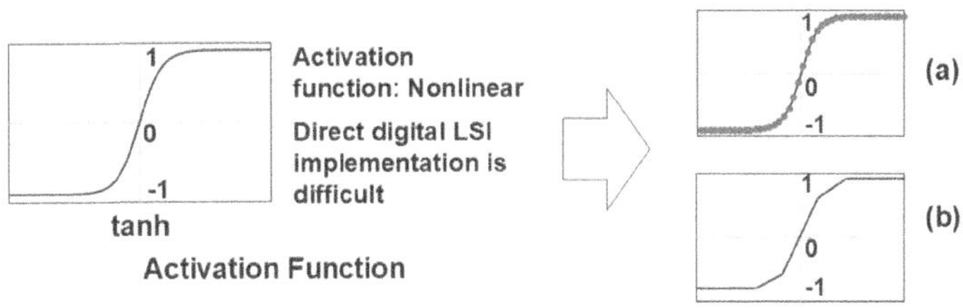

FIGURE 3.9
Activation function. (a) Lookup table (LUT) and (b) piecewise linear approximation (PWL).

by selecting the output value of the LUT corresponding to each input in the neural network. This method is generally widely used because it is fast, accurate, and easy to implement in integrated circuits. However, the size of the LUT can be large. In addition, if the accuracy of the LUT becomes low, the accuracy of the neural network may be adversely affected.

3.3 Analog Circuit Block

Analog circuits have a long history of investigation: in the 1980s, Carver Mead's group at the California Institute of Technology conducted pioneering research [4] to realize the function of neurons using analog circuits. Attempts were made to simulate information processing in sensory organs such as the silicon retina and silicon cochlea using LSIs.

Even today, in the era of deep neural networks, analog circuits are effective for processing in circuits with a small footprint. Moreover, since experience is required for the design and implementation of analog circuits, once a highly functional and high-performance neural network circuit is realized, it cannot be easily followed by others, and its superiority is high. As a basic structure, to realize an adder in an analog circuit, an op-amp adder or transistor adder is constructed using an op-amp or transistor. In an op-amp adder, the output is obtained by inputting signals to the inverting and noninverting inputs of the op-amp. On the other hand, there are many attempts to construct an adder at the transistor level, and there are examples of half adders and full adders constructed by utilizing the functions of MOS transistors and combining them to realize a high-speed, high-precision adder.

3.3.1 Adder

The explanation here will seem simple, but it is also one of the fundamentals of in-memory computing, which will be discussed in later chapters. In in-memory computing, the input and output are digital signals, but internally, it is important to compute with analog quantities.

Here, first of all, considering current signals, as an analog circuit, we simply combine the wires: if the current signal I_i as shown in Figure 3.10a is generated from each circuit, when they are connected, the following addition is obtained from Kirchhoff's law.

$$I_{\mathrm{OUT}} = \sum_{i=1}^{n} I_i \tag{3.1}$$

Also, if we consider charge as a signal, this can be done by electrically connecting capacitors.

$$V_{\mathrm{OUT}} = \sum_{i=1}^{n} V_i C_i \Big/ \left(C_O + \sum_{i=1}^{n} C_i \right) \tag{3.2}$$

Here, in in-memory computing, etc., C_i is assumed to be all equal, and if this is set to C as shown in Figure 3.10c, the following additions are obtained.

$$V_{\mathrm{OUT}} = \frac{C}{C_O + nC} \sum_{i=1}^{n} V_i \tag{3.3}$$

In these, current and voltage can be considered both positive and negative.

The method of converting an analog signal into a pulse signal train using a circuit that imitates a neuron and then adding this pulse signal train is also classified as an analog method, including the case where the signal is expressed in terms of the density of pulses.

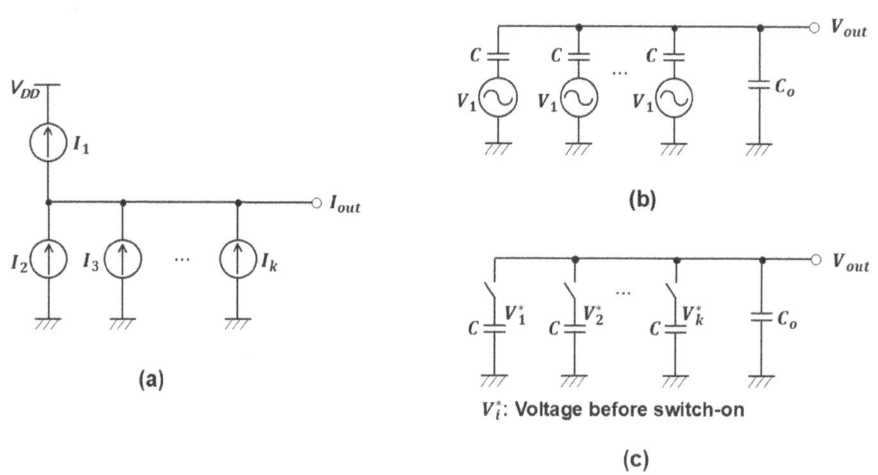

FIGURE 3.10
Adder. (a) AC current signal, (b) AC voltage signal, and (c) voltage signal addition by switch.

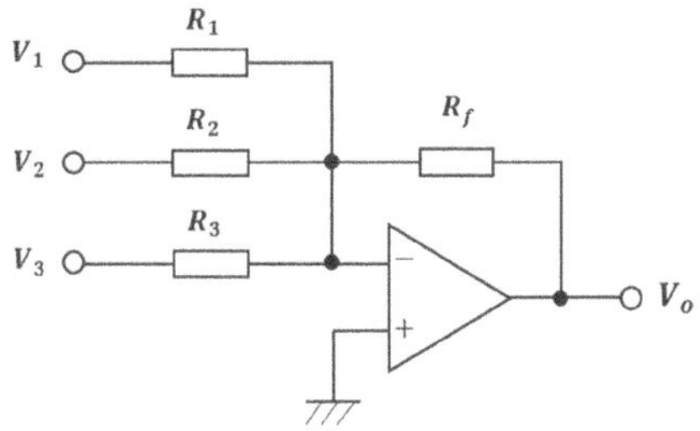

FIGURE 3.11
Multiplier.

3.3.2 Multiplier

In the study of imitating neural circuits, resistor networks, and in-memory computing, multiplication is performed using Ohm's law. That is, for example, if the magnitude of the aforementioned weights is given by the resistor R and the input signal is the current I, the voltage drop V is the result of multiplication, since the voltage drop is the following.

$$V = RI \tag{3.4}$$

From this and the aforementioned adder configuration, it is possible to perform operations that mimic the operation of a neural circuit using only passive elements. This was widely studied in the 1990s, for example, in operations that mimic the optic nerve [5,6]. This behavior was again exploited in the 2010s, as in-memory computing (discussed in Chapter 10).

In addition, as shown in Figure 3.11, a circuit that performs multiplication and addition/subtraction of weights and inputs can be configured by combining it with an OP amplifier. In other words, the magnitude of the resistor provided in the input section of the addition circuit using the OP amplifier becomes the weight, and the signal multiplied by the input signal to it is added to the OP amplifier.

3.3.3 Memory Circuit

In the digital circuit block, the register stored the data of the weights. In analog circuit blocks, these data are expressed in terms of resistance or the magnitude of charge. In other words, variable resistors or stored charge in capacitors are used. Some variable resistors can be realized by using memory elements that store analog information, as described in Chapter 10. The amount of charge stored in a capacitor can be used as the analog signal storage information as it is.

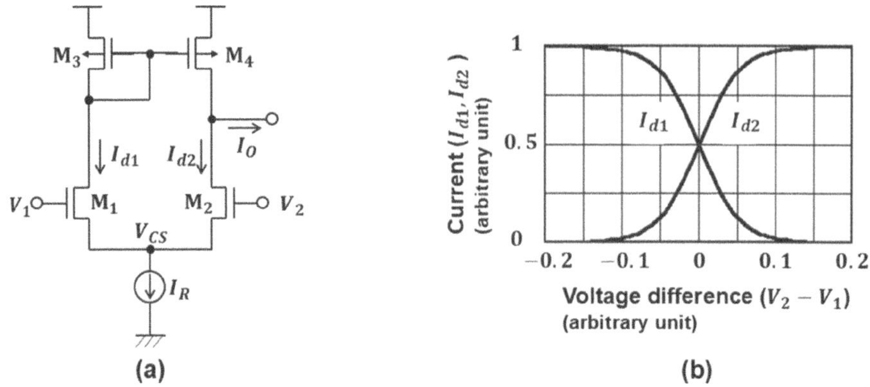

FIGURE 3.12
Nonlinear function. (a) MOS differential amplifier circuit and (b) input/output characteristics.

3.3.4 Nonlinear Conversion Circuits

In analog circuits, there is a method [4] of conversion that uses analog circuit elements such as operational amplifiers and transistors and takes advantage of the fact that their output characteristics are Tanh's characteristics with respect to the input. The method is shown in Figure 3.12, where two MOS transistors are used to create a differential pair by connecting them with their sources in common, and the subthreshold characteristics of the MOS transistors are utilized. The output current is given by the following equation, assuming that the MOS transistors are in the subthreshold region.

$$I_O = I_R \tanh \frac{\lambda \Delta V_{gs}}{2} \tag{3.5}$$

In this way, the tanh function can be realized. However, the characteristics of the subthreshold region will vary to some extent for each MOS transistor, even if they are placed in close proximity on the layout. Consideration must be given to this variation. In addition, it should be noted that the characteristics change significantly due to temperature change, which may affect the accuracy.

On the other hand, this circuit can also be used to realize derivative functions. The ability to realize such a distinctive function is the strength of analog circuits.

3.3.5 Use of Time Direction

The accuracy of analog circuits deteriorates when the supply voltage is reduced. However, in applications where there is a margin of time, analog signals can be processed with high precision by using a circuit that converts analog input signals to digital data [7]. For example, nonlinear conversion can be achieved by using a memory element to perform the integration of electric charge or current. A similar technique can represent the firing of a neuron by integrating electric charges and currents, and obtain an output signal according to the timing of the firing. The advantage of this method is that high accuracy can be achieved even with analog circuits. On the other hand, the calculation process is slower. The specific circuit is omitted, but reducing the mounting area is an issue, and it is necessary to devise ways to reduce the mounting area in the circuit that converts analog input signals to digital data and the circuit that performs the inverse conversion.

3.4 Low-Voltage and Low-Power Circuits

Low power is especially important when utilizing artificial intelligence LSIs at the edge. This section describes circuit or circuit block level low-power technology [8].

3.4.1 Principle of Low Power Consumption

Power consumption of large-scale integrated circuits (LSIs) is mainly composed of two components. One is dynamic power consumption, and the other is leakage current.

The dynamic power consumption of a CMOS circuit is as follows: f is the operating frequency (clock frequency), V is the supply voltage, C is the load capacitance that the CMOS circuit charges and discharges during operation, and α is the utilization rate of this circuit in overall LSI operation.

$$\text{Dynamic power consumption} = \alpha fCV^2 \tag{3.6}$$

Therefore, to achieve low dynamic power consumption, these values should be reduced. On the other hand, leakage current is the current that flows even when the LSI is not operating, and its low power consumption is discussed in the next section. It should be noted that leakage current flows even during LSI operation.

3.4.2 Clock Frequency Control

To lower dynamic power consumption, the main technique used is to reduce each α, f, V. α and f are handled collectively here.

First, dynamic power consumption can be reduced by turning off the clock signal when a particular CMOS circuit block is not needed. For this purpose, as shown in Figure 3.13a and b, a signal Enable is provided to select the target CMOS circuit block, and the clock is supplied only when this signal is output.

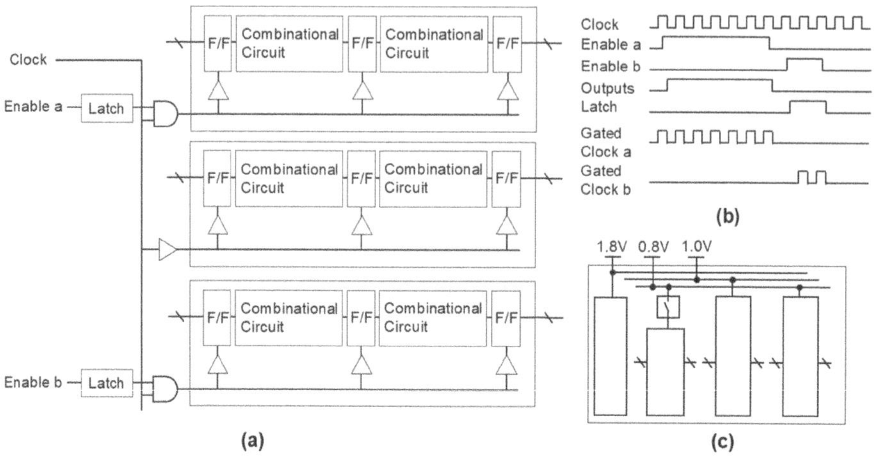

FIGURE 3.13
Low power circuits. (a) Clock gating structure, (b) timing chart of clock gating, and (c) multiple power supply.

This is called a gated clock. At this time, if the CMOS circuit block to be controlled is too small, the overhead of the control circuit is large, and the effect is small. In addition, clocks are usually structured as a clock tree with buffer circuits so that there is no difference in delay between internal circuits, such as the CPU, depending on the location. For this reason, a method is used to stop the clocks in this tree structure that do not need to operate in the circuit blocks connected to each branch. In artificial intelligence, although many sum-of-products circuits operate in parallel, it is still possible to control the clock supply for each circuit block corresponding to the layered structure of the neural network.

On the other hand, there is a large block of actions that should be called the mode of operation. In particular, since artificial intelligence processing is data-driven, its operation is different when it is capturing data and when it is processing data. Changing the value of the frequency for each of these modes of operation is effective in reducing power consumption. While data capture must be fast because the number of bits that can be processed at one time is small, data processing can be done in parallel, and therefore, stable operation can be achieved by lowering the frequency and increasing the degree of parallelism, which results in lower power consumption.

3.4.3 Voltage Control

Since dynamic power consumption is proportional to the square of the supply voltage, power consumption can be greatly reduced by lowering the supply voltage. This is already widely used in CPUs and general-purpose ASICs. However, the voltage of I/O circuits must match interface specifications, and analog circuits become unstable at low voltages. For this reason, as shown in Figure 3.13c, the power supply is divided into multiple areas with appropriate voltages set for each. Additionally, a switch circuit is sometimes inserted between the power supply and the circuit to implement a power disconnection design. Note that the time constant of the power supply lines is large and that the power supply voltage must be stable once set. Therefore, a mechanism is necessary to prevent malfunctions between power supply regions during power-up.

3.5 Low Leakage Current Circuits

Leakage current occurs because the transistor is not completely turned off. In particular, as miniaturization progresses, the transistor's gate oxide film becomes thinner and the leakage current increases. The increase in gate leakage current, explained below, is due to the fact that the insulating film (previously the oxide film) had to become thinner. The increase in subthreshold and GIDL currents is due to the reduced controllability of gate voltage, and junction current is due to the presence of pn junctions. These became issues in the 2000s, and many countermeasures were devised. In terms of devices, high-dielectric-constant insulating films are being placed under the gate, and MOS transistors are shifting from two-dimensional structures to three-dimensional structures. In the 2000s, circuit-level countermeasures [9,10] were also taken, many of which are still effective today.

3.5.1 Leakage Current

Figure 3.14a shows the leakage current in a MOS transistor with a planar structure [11]. (1) is the subthreshold current, which is the current that flows between the source and drain of the MOS transistor when the gate voltage of the MOS transistor is below the threshold voltage and is considered to be zero, i.e., turned off as a transistor. In other words, it is the current that flows even though it is not operated as an LSI and is called leakage current. When the threshold voltage that turns on the MOS transistor is lowered as the supply voltage drops, this current increases as a function of the lowered voltage. Below the threshold voltage, there is no inversion layer formed anywhere from source to drain in the silicon layer immediately below the gate. However, the current that flows due to the diffusion of conduction electrons is exponentially dependent on the gate voltage below the threshold voltage. What makes this current even more troublesome is that it has a large temperature dependence due to this mechanism, varying by more than one order of magnitude within the temperature range at which the LSI is guaranteed to operate. It is highly temperature-sensitive, so much so that it can be used to construct temperature sensors. (2) is gate leakage current, which is the current that flows through the insulating film between the gate electrode and the silicon layer. The thinner this insulating film is, the better the drain current can be controlled by the gate

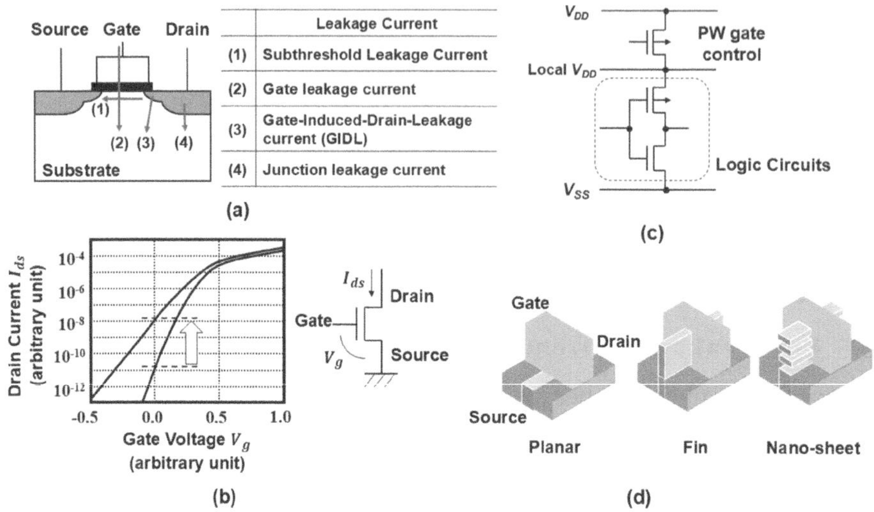

FIGURE 3.14
Leakage current reduction. (a) Leakage current of fine MOS transistors, (b) gated power, (c) subthreshold leakage current, and (d) 3D MOS transistors.

voltage, so the thinnest transistors have been developed along with the miniaturization of the transistors. For example, in a transistor with a gate length of 90nm, the thickness of this gate insulating film is 1.2nm if no measures are taken. This is the area where current can flow beyond the insulator due to the tunnelling effect. (3) is a leakage current called gate-induced drain leakage (GIDL). This occurs due to a mechanism that becomes particularly problematic when a reverse bias is applied to the drain and gate, which should be the off state. Because the depletion layer under the drain is thinner, if the electric field is higher here, the electrons will gain higher energy, and the current will flow beyond this. This is a mechanism that is not solved even with a three-dimensional structure. (4) is the leakage current caused by minority carriers in the pn junction. In logic circuits, this contribution is small.

3.5.2 Low Leakage Current Circuits

Many developments were made in the 2000s when the leakage current problem became apparent [9,10]. In this section, examples at the circuit or circuit block level are mainly described.

The first issue was the increase in subthreshold current due to the need to lower the power supply voltage in conjunction with MOS transistor miniaturization, and the threshold voltage was lowered accordingly. In LSIs using MOS transistors with a gate length of 65nm, this leakage current was said to be as large as the charge–discharge current during operation.

Figure 3.14b shows the gate voltage dependence of the MOS transistor drain current. While the gate voltage on the horizontal axis is linear, the drain current on the vertical axis is logarithmic. When the MOS transistor drain current is viewed on the logarithmic axis, it does not suddenly begin to flow at the specified threshold voltage when the gate voltage is varied. In this figure, the gate voltage is 0 V and the transistor is off, but current is flowing even here. If the threshold voltage is set higher, the current at the gate voltage of 0 V decreases exponentially, and this is not a problem. However, if the supply voltage is lowered and the threshold voltage is lowered accordingly, this current becomes large.

Figure 3.14c shows a method [12,13] that has been devised and widely used at the circuit level to reduce this subthreshold current in the entire LSI. The principle is simple: each circuit block has a switch between the power supply for that block and the power supply for the entire chip. This disconnects the power supply from the circuits that operate as a chip and from the circuits in the parts of the chip that are not in operation. In practice, this switch is also a MOS transistor, but one with a high threshold voltage is used here. This method, although simple in principle, was widely used because it was applicable to actual LSI logic circuit blocks and was highly effective. In addition, although not shown in the figure, a method was also employed to raise the threshold voltage by changing the voltage of the MOS transistor substrate of the nonoperating circuit block. Another issue is that the subthreshold current is temperature sensitive. Conversely, it can be reduced to a practically feasible value by lowering the temperature. For this reason, some designs solved this issue by combining the chip cooling method as well.

Finally, three-dimensional (3D) MOS transistors [14,15] should be mentioned. The three-dimensional structure improves gate control and reduces leakage current. In FinFETs, the gate surrounds the channel in three directions, making the channel narrower and enabling better field control by the gate voltage. The gate is surrounded on three directions of the channel, which makes the width of the channel narrower and allows better control of the electric field by the gate voltage. In nanosheet transistors, the channel is in the form of a thin sheet, and the gate completely surrounds the channel. This further enhances the electric field control and enables higher performance.

Regarding the reduction of gate leakage current, the use of high-k materials (high dielectric constant materials) is necessary, even as a three-dimensional structure. The same electrical effect can be achieved while increasing the thickness of the gate oxide film.

3.6 Summary

This chapter explained the basic electronic circuit blocks that are the building blocks of an artificial intelligence (AI) LSI. Electronic circuits include digital circuit blocks and analog circuit blocks. Digital circuit blocks include adders and multipliers, registers, latch circuits, nonlinear conversion processing circuits, and random number generation circuits. On the other hand, analog circuit blocks include circuits that perform addition and multiplication, memory elements, and nonlinear conversion circuits, as well as configurations that use time-domain signal processing. In addition, reducing power consumption is an important issue for AI chips. Minimizing leakage current is also becoming increasingly important in the design of low-power circuits as processes become finer.

References

1. S. Sedra and K. C. Smith, *Microelectronic Circuits*, 8th ed. Oxford: Oxford University Press, 2019.
2. D. Harris and S. Harris, *Digital Design and Computer Architecture*, 2nd ed. San Francisco, CA: Morgan Kaufmann, 2012.
3. C. Chapra and R. P. Canale, *Numerical Methods for Engineers*, 7th ed. New York: McGraw-Hill, 2015.
4. C. Mead, *Analog VLSI and Neural Systems*, Boston, MA: Addison-Wesley, 1989.
5. M. A. Mahowald and C. Mead, "The Silicon Retina," Scientific American Vol. 264 No. 5, pp. 76–82, 1991. DOI: :10.1038/scientificamerican0591-76
6. M. Carver and M. A. Mahowald. "A silicon model of early visual processing," Neural Networks, Volume 1, Issue 1, pp. 91–97, 1988. DOI: 10.1016/0893-6080(88)90024-X
7. K. Asada, T. Nakura, T. Iizuka, M. Ikeda, "Time-domain approach for analog circuits in deep sub-micron LSI," IEICE Electronics Express, Volume 15, Issue 6, Pages 20182001, 2018. DOI: 10.1587/elex.15.20182001
8. A. Chandrakasan and J. M. Rabaey, Low Power Digital CMOS Design, New York, NY, USA: Springer, 2015.
9. K. Itoh, "A Historical Review of Low-Power, Low-Voltage Digital MOS Circuits Development," *IEEE Solid-State Circuits Magazine*, vol. 5, no. 1, pp. 27–39, 2013. https://doi.org/10.1109/MSSC.2012.2230833
10. Y. Nakagome, M. Horiguchi, T. Kawahara and K. Itoh, "Review and Future Prospects of Low-Voltage RAM Circuits," *IBM Journal of Research and Development*, vol. 47, no. 5.6, pp. 525–552, 2003. https://doi.org/10.1147/rd.475.0525
11. K. Osada, "Reviews and Prospects of Nanoscale SRAMs," *2006 IEEE International Conference on IC Design and Technology*, Padua, Italy, 2006, pp. 1–8. https://doi.org/10.1109/ICICDT.2006.220835
12. T. Kawahara, M. Horiguchi, Y. Kawajiri, G. Kitsukawa, T. Kure and M. Aoki, "Subthreshold Current Reduction for Decoded-Driver by Self-Reverse Biasing (DRAMs)," *IEEE Journal of Solid-State Circuits*, vol. 28, no. 11, pp. 1136–1144, 1993. https://doi.org/10.1109/4.245594
13. T. Sakata, K. Itoh, M. Horiguchi and M. Aoki, "Subthreshold-Current Reduction Circuits for Multi-Gigabit DRAM's," *IEEE Journal of Solid-State Circuits*, vol. 29, no. 7, pp. 761–769, 1994. https://doi.org/10.1109/4.303713

14. R. S. Pal, S. Sharma and S. Dasgupta, "Recent trend of FinFET devices and its challenges: A review," 2017 Conference on Emerging Devices and Smart Systems (ICEDSS), Mallasamudram, India, pp. 150–154, 2017, DOI: 10.1109/ICEDSS.2017.8073675
15. N. Loubet et al., "Stacked nanosheet gate-all-around transistor to enable scaling beyond FinFET," 2017 Symposium on VLSI Technology, Kyoto, Japan, pp. 230–231, 2017, DOI: 10.23919/VLSIT.2017.7998183

4

Basic Structure of AI Chips and Various Neural Networks

In the previous chapter, the components and basic electronic circuits of an artificial intelligence LSI were presented. In this chapter, we will look at the basic structure of the artificial intelligence LSI to be built on this basis. First, the structure of a neural network (NN) is shown, and the LSI circuit configuration is explained using a simple three-layer structure as an example.

4.1 Configuration of Neural Network Inference Function LSI

4.1.1 Representation Power of Neural Networks

In a NN, there are many artificial neurons (nodes) in each layer that make up multiple layers [1]. Each neuron weights the input signal and calculates the output using an activation function. NNs with this structure are known to be able to represent a variety of functions. Here, a function is a relationship between inputs and outputs. By adjusting the weights and biases between the layers of the NN through learning, the relationship between inputs and outputs can be freely expressed.

As shown in Figure 4.1, taking handwritten character recognition as an example, this is the relationship between the image data of the handwritten characters, which is the input x_i, and the character data corresponding to the output z_i. It is a function that obtains its output from such an input. Specifically, in this recognition example, the input is the image data of the handwritten characters, divided into 28 x 28 rectangles and the shading in each rectangle is binarized here to 784 binary data. The output corresponds to 10 numbers 0 ~ 9. If we represent output 0 as 1000000000 and output 1 as 0100000000 (a form we call one-hot representation), this is 10 binary data. This relationship, or function, between the input x_i and the output z_i can be realized by a NN, which is artificial intelligence. A NN consists of multiple layers, and each layer has multiple neurons. This structure can represent a complex function and obtain a desired output for an input. To realize this function, the NN learns weights using teacher data. In other words, in the recognition of handwritten characters, a large amount of handwritten character data and the corresponding correct answer label (0–9) are given as

DOI: 10.1201/9781003599067-4

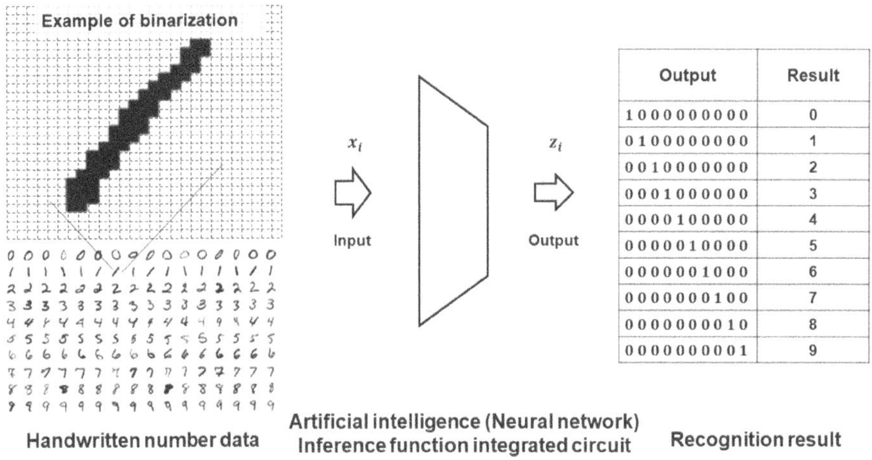

FIGURE 4.1
Neural network function – Inference.

training data, and the NN adjusts the weights using these data. In this way, it learns the relationship between input and output. At this time, in the NN, the use of nonlinear activation functions allows for a rich representation of the characteristics of the data, i.e., the relationship between input and output, which can be expressed as a function. Although it is not a functional form that can be expressed by a formula, it is possible to realize a function that is the relationship between input and output by using a relatively simple structure of weights, network, and activation function.

By converting this NN into an LSI, high-speed, low-power, and low-cost processing is possible. Let us first try to assemble an actual NN with a conceptual circuit block.

4.1.2 Basic LSI Circuitry for Inference Functions

Let's try to construct a NN capable of performing arbitrary functions using LSI circuits. As mentioned earlier, the circuit (neuron) of the NN consists of the sum-of-products operation and the activation function. Therefore, it is sufficient to prepare and connect these components.

The overall structure of the integrated circuit is first shown in Figure 4.2. The number n of neurons in each layer depends on the artificial intelligence to be configured. In the present example, the input layer is simply buffers that store the data x_1 ,... x_k ,.. x_n. For the image data of handwritten characters, the image data is divided into 28×28 rectangular regions (784 in total). The data consist of the shades of each area converted into binary values, and 784 buffers that store this binarized data are lined up. The next layer is the intermediate layer (also called the hidden layer), which is composed of 256 artificial neuron circuits. (The optimal number of neurons is not discussed

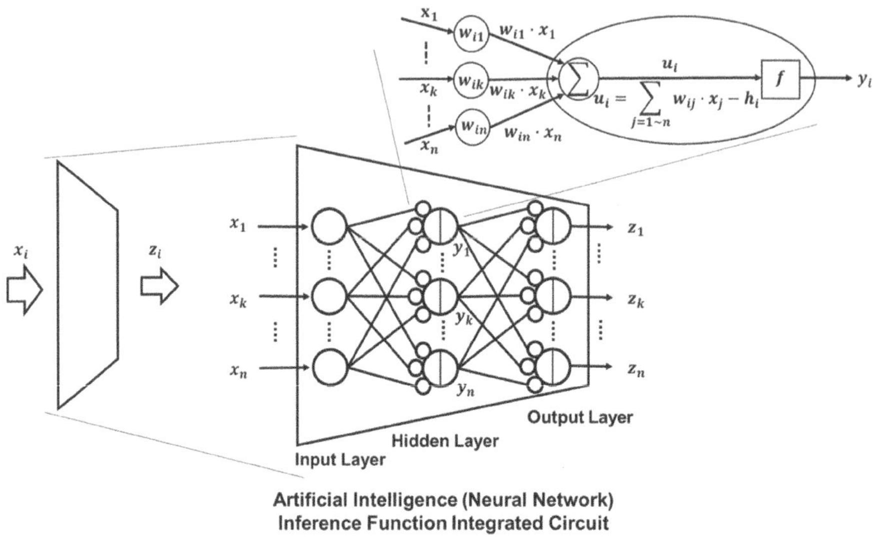

FIGURE 4.2
Basic configuration of neural network.

here, but will be determined separately in relation to the required recognition accuracy (the ratio of the probability that a handwritten character image can be recognized as a correct digit)).

Then, each output of one buffer in the input layer must be connected to 256 artificial neuron circuits in the middle layer. Also, each of the 256 artificial neuron circuits in the middle layer would have 784 inputs. The output layer consists of 10 artificial neuron circuits, and these inputs are connected to the outputs of each of the 256 artificial neuron circuits in the middle layer. In other words, the number of connections is extremely large, and the LSI needs to implement this huge number of connections.

Thus, the content to be kept in mind in the LSI of NNs, along with the characteristics of the aforementioned unit circuit itself, that is,

- There are many similar unit circuits,
- These are densely connected.

NNs themselves have a long history of research: the second boom in research and development of artificial intelligence occurred in the 1990s, and integrated circuits were also considered [2]. The difference between that time and the third boom from the 2010s is the substantial development of miniaturization technology and embedded memory device technology. Miniaturization technology has become two to three orders of magnitude finer, and the use of novel nonvolatile memories, as will be discussed in Chapter 9, is flourishing. In NNs, where there are many similar unit circuits and these are densely connected, it has become easier to mount large-scale networks on LSI chips.

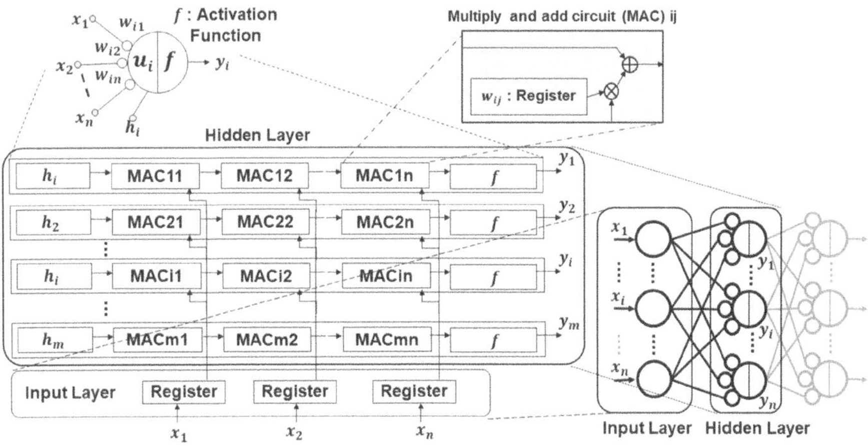

FIGURE 4.3
Simplified Example for constructing neural networks.

Let us discuss the realization of this dense connection a little more concretely with a simplified structure. This would be a two-dimensional array structure as shown in Figure 4.3. This structure can realize one layer of a NN. It consists of a buffer (register) of inputs arranged one-dimensionally in the x-direction (the horizontal direction of the paper), a circuit storing weights and operations arranged in an array, and an activation function arranged one-dimensionally in the y-direction (the vertical direction of the paper).

First, the inputx_i is stored in a register located in one dimension in the X direction at the bottom of Figure 4.3. This is the input layer shown in the lower right of the same figure, or in Figure 4.2, which is simply a buffer circuit to store data. Next, a simplified sum-of-products circuit ij is placed in the form of an array. This is a cluster of artificial neuron circuits, corresponding to the middle layer in Figure 4.2. The simplified sum-of-products circuit ij is the circuit that stores the weights w_{ij} and the necessary operations. These circuits are connected in the X direction, and each has sum-of-products circuit ij has two inputs and one output. The first input is the signal from the buffer of inputs arranged in one dimension above, and the second input is the output from the previous stage that makes up the row. In each circuit, the product of the signal from the first input and the stored weights is taken, and this is summed with the signal from the second input. This is computing the sum of products. Thus, in this one line, n sums of products can be performed. The arrangement consists of m rows of these elements, with circuits designed to perform activation function operations positioned for each row. This completes the arrangement for one layer.

Let's look at the above again, in a slightly sequential manner, in Figure 4.4. First is the input layer shown on the left in (a). The input layer consists of n buffers that store, $x_1 ... x_i$, $... x_n$. Each of which has a latch circuit as shown on

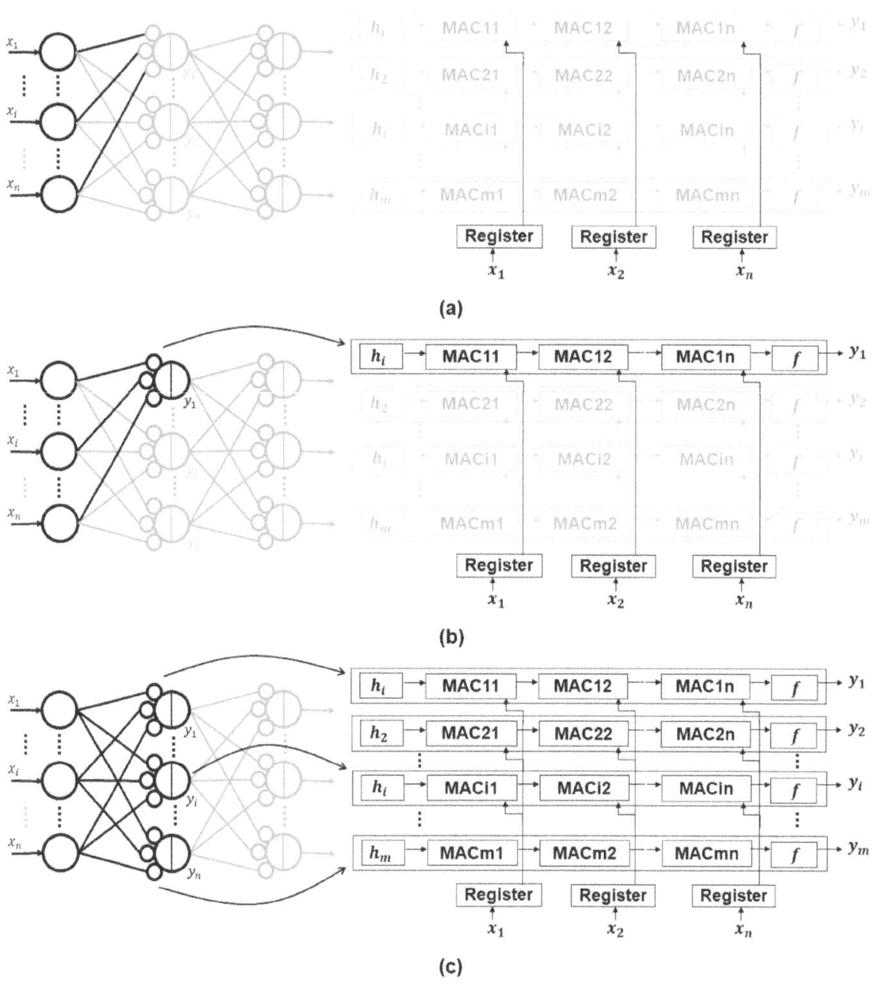

FIGURE 4.4
Correspondence between neural network and circuit configuration in one layer. (a) The relationship between the input layer of neurons and the one-dimensional buffer (register), (b) The relationship between a single neuron in the intermediate (hidden) layer and the circuit layout. Each circuit multiplies the input from the register by the stored weight and adds it to the signal from the previous stage's output, (c) The relationship between all neurons in the intermediate layer and the circuit layout. All operations can be performed in parallel.

the right and stores the value of the latch circuit. The outputs of these latch circuits are then directed upward to the middle layer, the sum-of-products circuit *ij*, as shown in this figure.

Next, Figure 4.4b considers an artificial neuron circuit in one of the intermediate layers. Here, n data from the registers are input in parallel to n sum-of-products circuits in MAC1j ($j = 1 \sim n$). Each of the sum-of-products circuits in MAC1j contains the weights w_{1j}. Each circuit takes the product of the input from the register and the stored weights and sums it with the signal from the previous output. This means that the sum of products of one artificial neuron is calculated. The calculation of $\sum_{j=1}^{n} w_{1j} \cdot x_j - h_1$ is done in this line, considering also the $-h_1$. The result is input to the activation function f, which yields y_1 as its output.

As shown in Figure 4.4c, there are m lines. The demand here is that all of the operations in this m line can operate in parallel. Therefore, this circuit block can perform the operations in the following equation, which is a restatement of the equation described in Chapter 2.

$$u_i = \sum_{j=1}^{n} w_{ij} \cdot x_j - h_i \tag{4.1}$$

$$y_i = f(u_i) \tag{4.2}$$

In this figure, dedicated circuits are arranged for each line of the activation function, but the circuit block that performs this operation can be configured as a single circuit block that performs each line of the activation function in turn. Although the activation function part may require a relatively large circuit configuration, this configuration can reduce the mounting area. The above operations may be synchronized with the clock, or the entire operation may be performed asynchronously.

With this structure, a single-layer NN can be implemented on an LSI. If we further prepare circuit blocks with a similar structure, a multilayered structure can be realized. This circuit can process a large amount of inputs because the u_i and y_i operations are processed in parallel.

Furthermore, let us look again at the NN configuration for the recognition of handwritten numbers.

First, prepare the input data as shown in Figure 4.5. The input is a handwritten digit image data set [3] from 0 to 9, known as MNIST. Each image data set is 28 × 28 pixels (=784 pixels) in size, and each pixel of MNIST data is 8 bits, but here, as already mentioned, each pixel is assumed to be binary (1 bit) with reduced precision (right side of the same figure). As shown in the upper part of the figure, this is rearranged into a sequence of 784 binary values $x_i (i = 1 \sim 784)$. This is the input. The first image data is two-dimensional, and we recognize numbers visually from it, but for a NN, reordering this is simply an input, and it is sufficient to obtain the relationship between this and the output. Therefore, the reordering can be arbitrary as long as it is the same for all data. However, there is also a method that takes advantage of the characteristics of the two-dimensional array. In this case, the handling is appropriate for that method.

This is input to the NN from the left side in the same figure, as shown in Figure 4.6. The output on the right side would be the data corresponding to the numbers from 0 to 9. Here, using the subscript j, since the number of

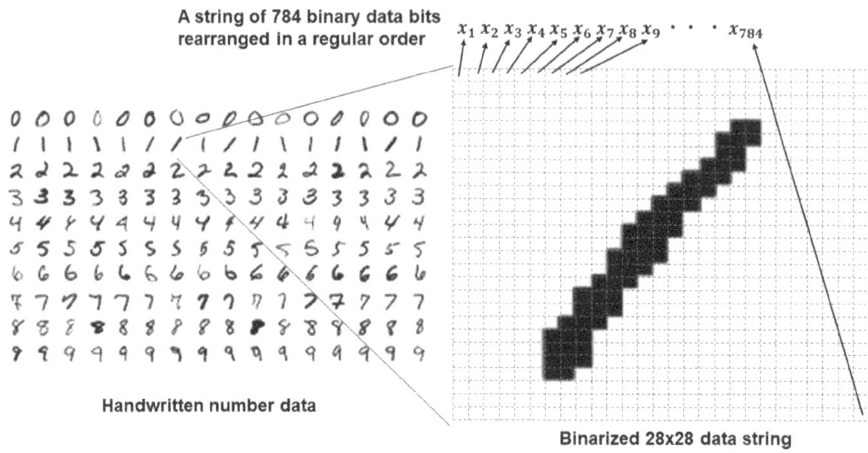

FIGURE 4.5
Example of input data preparation for handwritten character recognition.

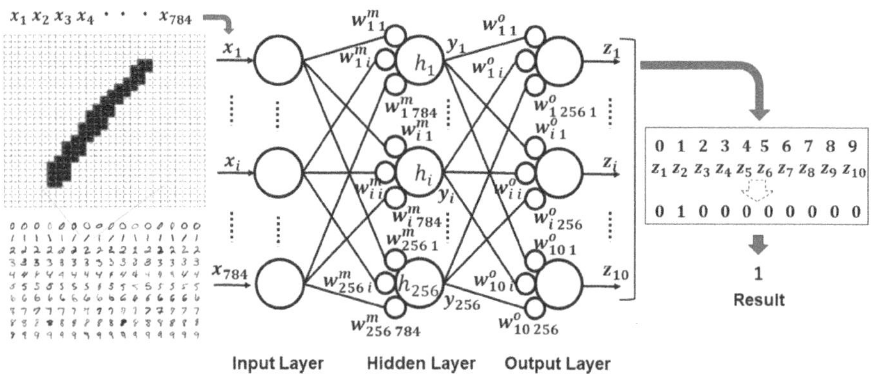

FIGURE 4.6
Construction of character recognition neural network.

pieces is different from the input, we get $z_j (j = 1 \sim 10)$. This is the data from $z_1, z_2, \cdots, z_{10}$ each of which is a sequence of 0s or 1s, which corresponds to the numbers 0–9. For the converted handwriting given as input, the specific output representing that number will be 1, and the other output will be 0.

At this time, there are multiple handwritten character image data (16 in this image), as shown in Figure 4.5 left or Figure 4.6 lower left, even if they are the same "1" for example, and the weight w_{ij} is adjusted to be the same for them. This is learning. This method results in a NN that can recognize handwritten characters. This entire NN is a function that can give the output $z_j (j = 1 \sim 10)$ for an extremely large number of pairs of inputs $x_i (i = 1 \sim 784)$. Furthermore, it becomes feasible to identify data other than that which was prepared for the training process (i.e., generalization ability).

This is implemented as a circuit for the LSI, and the main parts are the same as those described in Figure 4.3. However, the example in Figure 4.6 has 784 inputs. On the other hand, the intermediate layer has 256 inputs and 10 outputs.

Therefore, the number of input registers is 784. In the two-dimensional array shown in Figure 4.6, where sum-of-products is performed in the middle layer, there are 784 sum-of-products circuits in the X direction and 256 circuits in the Y direction, each of which has an activation function circuit in the final stage. The 256 outputs, which are the pairs of activation functions, are the inputs to the final stage.

The output layer in Figure 4.6 has 10 outputs $z_j(j = 1 \sim 10)$ for 256 outputs. Since the operation of sum-of-products is similar, there are 256 input registers and 256×10 sum-of-products circuits beyond them. Based on this result, the output $z_j(j = 1 \sim 10)$ in the example in Figure 3.6 is determined to be 1 because of the sequence 0100000000. Note that a circuit that realizes the softmax function may also be used, but is omitted from this explanation.

Up to this point, we have taken the example of character recognition to see how a NN can be constructed as an integrated circuit in a simplified configuration. The example here is intended for application to in-memory computing, which will be discussed in Chapter 10.

4.2 Configuration of Neural Network Learning Function LSI

NNs can "learn" to adapt to given input data and produce desired outputs. However, learning requires a large amount of data and is a process that requires a large amount of computational resources. It is not reasonable at this time to implement this process in a closed form in an LSI. In general, learning is performed separately, and a function that stores the weight data obtained from the learning process and performs inference processing is installed. However, it is useful to reduce power consumption and increase speed by specializing in the content of individual processing at the edge and incorporating some learning. Future development of LSIs for learning functions, including these functions, is expected in in-memory computing, as shown in Chapter 10. In this section, we will look at the basic functions of learning in general, assuming that LSIs will be used in the future.

The LSI circuits required for learning are circuits that determine the amount of internal weight updates for the artificial neuron elements and circuits that rewrite the contents of the memory elements that store the internal weights. As a function, it must be able to calculate the error during learning and update the weights afterward. In the error back propagation method described in the next section, the derivatives of the activation function are also required for the error calculation. The Boltzmann machine learning algorithm, on the other hand, uses a simulated annealing technique. In this algorithm, the weight values are adjusted by applying a temperature parameter to the Boltzmann distribution to determine the slope of the probability function.

4.2.1 Neural Network Training

Learning is the process of updating the weights, including the bias, to give the desired output for a given input. In this section, we will discuss the basics of error back propagation (BP) [4], which is widely used as a learning algorithm.

For illustration, take supervised learning as an example in the NN shown in Figure 4.7. For a given input, let T_i be the value of the signal we want from the teacher data, and let y_i^n be the output of the nth layer as the NN is multilayered. The weights show this in the superscript to indicate from which layer to which layer it is at the time. In the example in Figure 4.7, the third layer is the output, and to make this output y_i^3 closer to the teacher data T_i, we define an error function that is the difference. This function takes a smaller value when the two are close (e.g., the squared error function, which will be used later in this explanation), and learning is to modify the weights so that this function is minimized. In other words, as shown in Figure 4.8, the slope (gradient) is calculated by differentiating the weights with respect to the error function to be minimized, and the value of the error function is reduced by adjusting the weights in the opposite direction of the slope. The calculation of the gradient is thus required, which is done using the derivatives of the activation function, as shown below. There are two methods for updating the weights: the steepest descent method, which uses all the training data to update the

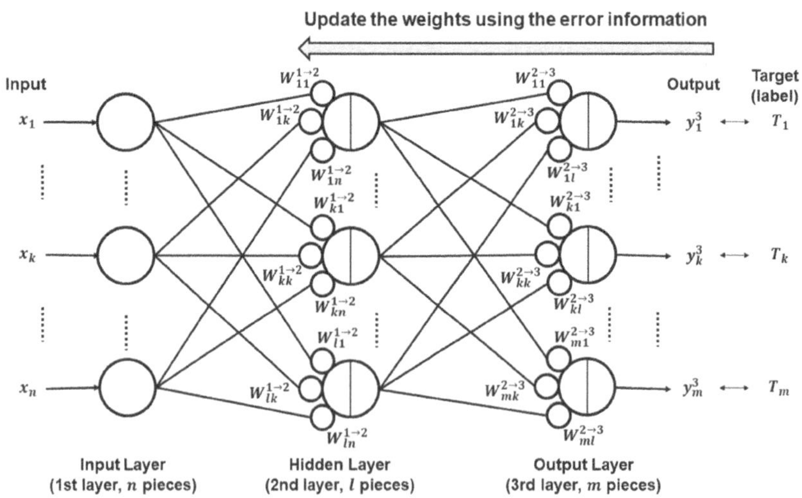

FIGURE 4.7
Neural network training.

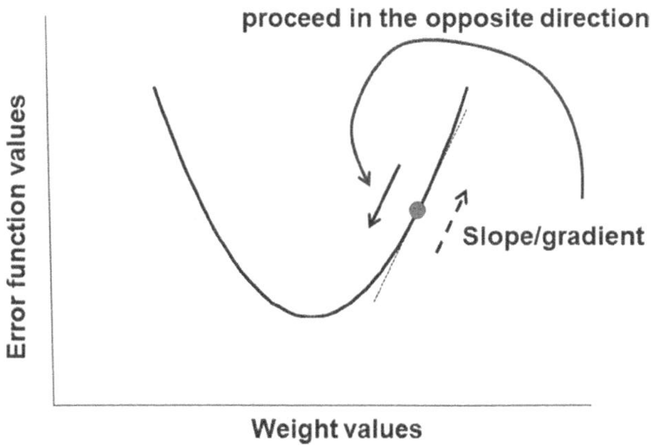

FIGURE 4.8
Weights update.

weights, and the stochastic gradient method, which uses some of the training data to update the weights. The latter can reduce the computational cost, but the learning progress may be unstable.

4.2.2 Error Back Propagation Method Overview

In the commonly used error backpropagation method, assuming that there is a multilayer NN, the weights of a layer (here, the kth layer) are first updated based on the error, which is the difference between the output of the layer and the desired result. Next, the same operation is performed for the $k-1$ layer. The weights are updated as this is done sequentially to the previous layer. In this way, the signal flow proceeds in the opposite direction compared to the cases described so far, a technique that minimizes the overall error. This process is repeated for each layer in turn, starting with the output layer, and when the input layer is reached, the process is repeated again, starting with the output layer.

Looking specifically at the output layer, there is explicit teacher data for training the NN in the output layer, but in order to proceed sequentially to the previous layer, each layer requires its own teacher data. This will be sent sequentially from the output layer as shown below. Therefore, we proceed with processing at each layer by assuming that there are m signal values of the training data propagated from the entire NN's training data T_i (m per learning). For this T_i, let k be the layer of the NN to be trained and y_i^k be its output. Taking the squared error function as the error function E, we get the following.

$$E = \frac{1}{2}\sum_{i=1}^{m}\left(T_i - y_i^k\right)^2 \tag{4.3}$$

Using this, as explained in Figure 4.8, the gradient is calculated by differentiating the weights with respect to the error function to be minimized, and the value of the error function is reduced by adjusting the weights w_{ij} in the opposite direction to the magnitude of the gradient. Therefore, the formula Δw_{ij} for updating the weights w_{ij} to be adjusted is as follows. Note that although defined between each layer, the superscript indicating this has been omitted.

$$\Delta w_{ij} = -\eta \frac{\partial E}{\partial w_{ij}} \tag{4.4}$$

Here η is called the learning rate, which is a coefficient that determines how much the gradient $\frac{\partial E}{\partial w_{ij}}$ is used, and it is also a value that indicates how fast learning occurs. Usually, a small value such as 0.1 is used. The minus sign indicates a change in the direction opposite to that of the gradient magnitude. On the other hand, y_i in equation (4.2) is a function of u_i and this u_i is a function of w_{ij}, so from the chain rule of differentiation,

$$\frac{\partial E}{\partial w_{ij}} = \frac{\partial E}{\partial y_i} \frac{\partial y_i}{\partial u_i} \frac{\partial u_i}{\partial w_{ij}} \tag{4.5}$$

It follows that the elements of the right-hand side are shown using equations (4.1) and (4,2):

$$\frac{\partial y_i}{\partial u_i} = \frac{\partial f(u_i)}{\partial u_i} = f'(u_i) \tag{4.6}$$

$$\frac{\partial u_i}{\partial w_{ij}} = x_j \tag{4.7}$$

x_j is an input to y_i, so it corresponds to the output of one previous layer when considering multiple layers. Δw_{ij} is the following equation.

$$\Delta w_{ij} = -\eta \frac{\partial E}{\partial y_i} f'(u_i) x_j \tag{4.8}$$

A similar relationship can be derived for the intermediate layers; only the chain rule of differentiation is longer. We can compute Δw_{ij} for each layer until we get to the input layer.

To achieve this, the following functions are required for the LSI circuit of the kth layer NN.

1. Calculate the slope (gradient) by differentiating the weight with respect to the error function to be minimized.
2. Adjust the weight in the direction opposite to the slope.

The calculation of this 1. requires the derivatives of the function. Many of these can be expressed using the activation function itself, as shown in Table 2.1 in Chapter 2. We build a circuit equivalent to this and use the output of this circuit to update the weights of the k layer.

What should be considered in the LSI circuit design is that, first, values can be calculated sequentially layer by layer from the input in forward propagation, and then the above calculations can be done in reverse direction from the output to the input, layer by layer. In the forward propagation, parallel computation was possible on a layer-by-layer basis, and in learning, parallel computation is possible by using the same configuration for updating the weights. Therefore, the new functionality required for learning here is the calculation of $f'(u_i)$ on the output side of each layer.

However, in learning, it is necessary to adjust the learning rate and determine the end of learning by looking at the size of the error and its change during the learning process. In addition, in many cases, a larger value is used for the significant digits (number of digits) of w_{ij} in learning than in inference. For this reason, the learning function is often limited when incorporated into the edge.

In the error back propagation method, learning is performed by back propagating minute changes in weights, which is considered to be a nonbiological calculation that differs from information processing in the brain. In addition, the error back propagation method adjusts the overall weights by back propagating errors, but it is believed that local learning occurs in the actual information processing of the brain. Furthermore, the error back propagation method assumes a dense representation in which all neurons are activated at the same time, which is an advantage of parallel processing, but it is believed that sparse representation is used in brain information processing due to its high energy–performance ratio. To address these issues, Hebbian learning for local learning and sparse coding for sparse representation, for example, have been proposed.

4.3 Summary of Basic AI Chip Configuration

This configuration can be used to construct a NN. What is important here as an LSI is that obtaining multiple outputs from multiple inputs can be realized with a two-dimensional array-like circuit as described in the configuration of the previous Figure 3.1. That is, the configuration is as shown in Figure 4.9. In NN operations, LSI circuits consist of a large number of processing elements arranged in a two-dimensional array, and their outputs are processed by activation functions, enabling two-dimensional calculations. This is a parallelized computation.

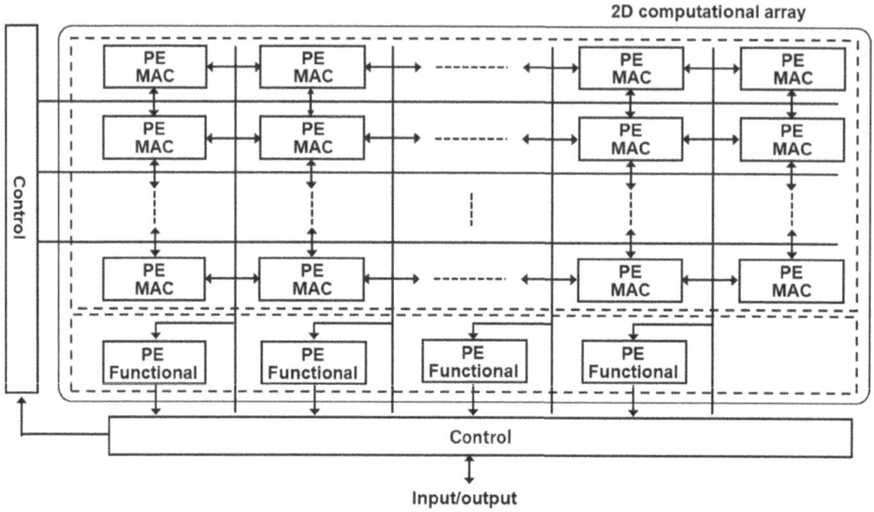

FIGURE 4.9
Major components of AI processing circuits.

Note that the presence of connections may vary depending on the specific NN implementation, not all connections are made. This is sometimes called sparse connections. This can also be easily realized with this structure, as long as the outputs from the input buffers in the *X* direction are not partially connected to the circuit on the array, and the inputs in this case can be fixed to either high or low potential (depending on the logic configuration). Sparse connections can be made meticulously.

Figure 4.10 shows the structure of a typical digital circuit that performs parallel processing [5], which is important in this study.

First, the basic configuration that uses digital circuits to perform arithmetic operations is called a processor element (PE). Figure 4.10a is an example. In artificial intelligence processing, as described in Chapter 1, a large amount of data is prepared in advance and processed all at once. Therefore, it is important to determine the necessary processing in the PE to be used and to have a configuration that operates many of them in parallel. For this purpose, single instruction multiple data (SIMD) shown in Figure 4.10b is a configuration in which a single instruction is executed on multiple processors or PEs to process multiple data in parallel. To further increase the parallelism, PEs are arranged in an array as shown in Figure 4.10c. LSIs for artificial intelligence processing often take the SIMD type or this array configuration. Here, the control circuit is used to exchange data between the PEs and the memory. Once the data is obtained from the memory, it is processed in parallel while propagating PE. In other words, for multiple arithmetic units, the output results of each are not returned to memory.

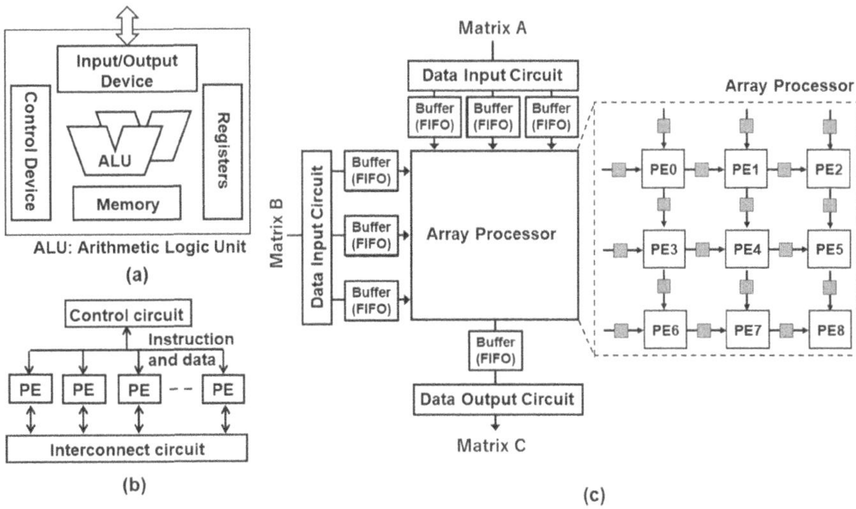

FIGURE 4.10
Digital circuit blocks. (a) PE configuration, (b) SIMD (single instruction, multiple data) structure, and (c) Matrix multiplier unit.

The processing results are collectively returned to memory. There are various types of NNs that constitute artificial intelligence processing, and NN computation is a collection of matrix operations of various sizes. By processing the results of many multiplications and additions using PE with an activation function, nonlinear behavior is obtained from the accumulation of many linear operations. The optimal configuration to realize this in hardware is shown in Figure 4.10c. GPUs, which can accelerate computing while specific processing in a spatial arrangement, also excel in parallel computing, but this requires the implementation of a variety of features needed in modern general-purpose processors. Just to list the terms of component and function terminology, there are many examples, such as cache, branch prediction, out-of-order execution, multithreading, speculative prefetching, address calculation, context switching, and so on, all of which improve the average performance of the processor. All of these improve the average performance of the processor, but are unnecessary for the specific application of artificial intelligence. The omission of these factors allows artificial intelligence chips to achieve high performance.

However, since the method of arranging PEs in an array requires knowledge of circuit design, a method in which the weight data is stored in a memory element as the first circuit block and the part that performs calculations separately from this as a second circuit block, while exchanging information as if these were performed by an ordinary memory and arithmetic unit. In the case of FPGAs, the configuration is often a mixture of the two.

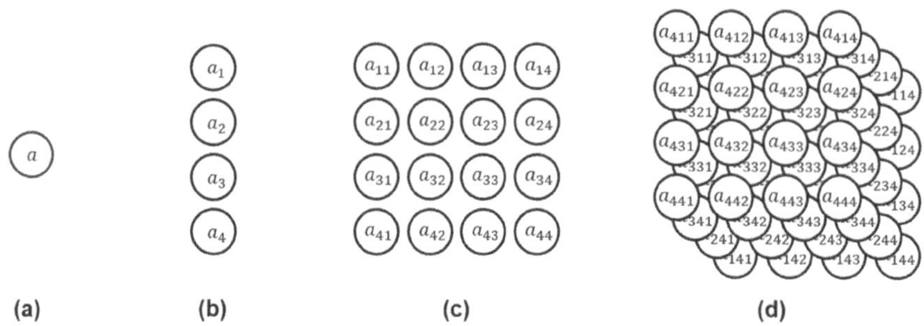

FIGURE 4.11
Scalar, vector, matrix, and tensor. (a) Scalar, (b) vector, (c) matrix, and (d) tensor.

At the end of this section, Figure 4.11 summarizes the types of data used in arranging numbers for artificial intelligence processing calculations. In other words, as seen in Equation 4.1, artificial intelligence processing was the following operations.

$$u_i = \sum_{j=1}^{n} w_{ij} \cdot x_j$$

Looking at this equation, the left side represents a single number and is called a scalar. On the right side, on the other hand, the calculation is performed between x_j and w_{ij}, which is the inner product of the vector x_j and the matrix w_{ij}. The operation of the artificial intelligence process was to perform an extremely large number of these calculations. As shown in Figure 4.12, a scalar is a single number, as shown in (a) as a, and a vector is a one-dimensional arrangement of numbers, as shown in (b). A matrix is a two-dimensional arrangement of numerical values as shown in (c). To express numerical values for each of these components, a vector is represented with a single subscript, such as a_i, $(i = 1 \sim 4)$, while a matrix is represented with two subscripts, such as a_{ij}, $(i = 1 \sim 4,\ j = 1 \sim 4)$. A higher dimension can be considered and is called a tensor; a three-dimensional one is shown in (d). The three-dimensional tensor is shown in (d) and is denoted by three subscripts, a_{ijk}, $(i = 1 \sim 4,\ j = 1 \sim 4,\ k = 1 \sim 4)$. An even higher-dimensional tensor can also be considered. A scalar can be regarded as a zero-dimensional tensor, and a vector is sometimes called a one-dimensional tensor.

Here, data is often more than three-dimensional in artificial intelligence processing. For example, an image is usually represented as a 3D tensor (height, width, and color channels). If this data can be processed as is at high speed, the calculation will be efficient. For this purpose, in the example, the data is expanded to 1D and weighted in 2D. However, if the initial data is a 3D tensor, calculation methods and LSI methods are being developed to calculate it as it is.

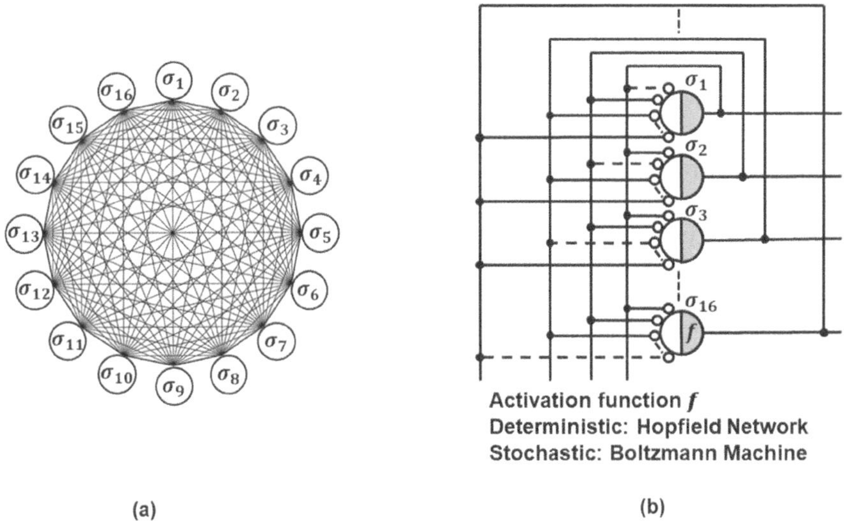

FIGURE 4.12
Convolutional neural network. (a) Fully connected neural network and (b) state transition method.

4.4 Neural Network Structure

By combining the above basic configurations, the basic structure of a NN circuit itself can be constructed. However, the structural design of a NN exhibits significant variability, contingent upon the specific function it is intended to fulfill. Based on these characteristics, it is necessary to devise a way to increase the processing capacity with lower power and a smaller mounting area.

This section provides an overview of NN [1] structures that have proven beneficial for various applications. Additionally, we will address issues identified in the implementation of LSI.

4.4.1 Feedforward Multilayer Network

Feedforward networks have been employed as illustrative examples in preceding sections. The model comprises an input layer, an output layer, and typically multiple intermediate layers, also referred to as hidden layers. The data necessary for the resolution of the problem is entered into the input layer. The input layer's function is to output the input data in its original form. Despite the composition of NNs into multiple layers, the input layer is occasionally excluded from the calculation and is regarded as the zero layer. In the hidden layers, the weighted sum of each component of the input data is input. This is then nonlinearly transformed using an activation function and output. In the output layer, the outputs from multiple hidden layers are integrated, and the requisite transformations are executed prior to output. In the context of classification problems, the softmax function is employed to facilitate this transformation.

As an LSI, it comprises a substantial array of fundamental arithmetic circuits interconnected in a layered network configuration, with signals propagating in a single direction from input to output. The computations within each layer are executed through the utilization of a two-dimensional calculation array. Consequently, advancements in microfabrication technology have enabled the integration of extensive networks on a singular chip. Moreover, endeavors are underway to curtail energy expenditure and augment processing speed by leveraging the characteristics of circuits and networks. The system's compatibility with in-memory computing, a subject to be addressed in subsequent chapters, is a notable feature.

4.4.2 Convolutional Neural Networks

Convolutional neural networks (CNNs) are a type of feedforward NN consisting of an input layer, an output layer, and multiple intermediate layers. The distinctive structural composition of the intermediate layers is illustrated in Figure 4.12a, which encompasses convolution layers and pooling layers. In the convolution layer, information from adjacent cells is aggregated. This facilitates the detection of patterns within images and is also an effective method for classifying various types of data, such as audio, video, and text, beyond just images.

Specifically, as shown in Figure 4.12b, multiple kernels are prepared for the input data. In this particular instance, the input data is represented by a 6×6 matrix, which is indicative of a two-dimensional image composed of 1s and 0s. A 3×3 matrix representing the kernel (filter layer) is shifted vertically and horizontally one by one over the input data, and the product of each element and component of the matrix is calculated. This calculation is implemented in an LSI by taking nine data points from the input data and the kernel and calculating their products in parallel. In this scenario, the input data undergoes a shift, as illustrated in the accompanying figure. However, it is imperative to note that all calculations are initially executed for the first kernel (1). Subsequently, in the pooling layer, the output is simplified by performing nonlinear downsampling. To illustrate, this process implements the extraction of the maximum value of the data within a specific domain as a singular data point. This approach ensures invariance to minor alterations in the input, such as shifts or rotations. This process enables the acquisition of features corresponding to the specified kernel. Subsequently, the aforementioned process is repeated for the subsequent kernel (2), thereby acquiring the features pertinent to this kernel.

In general, these steps are repeated dozens to hundreds of times, gradually extracting features. Upon receiving an input image, the initial layers determine low-level features such as edges, rectangles, and arcs. Subsequent layers identify more complex intermediate-level features, such as geometric shapes and simple objects. Later layers determine high-level features. Although processing is predominantly executed in the cloud, endeavors are underway to enhance filter sizes, discontinue superfluous calculations, and repurpose calculation results to leverage AI chips for edge processing.

4.4.3 Hopfield Networks and Boltzmann Machines

A Hopfield network [6] is an asynchronous network with symmetric interactions between neurons, as shown in Figure 4.13a. It is a deterministic NN that generates outputs by leveraging the network's capacity to minimize its energy based on relationships that have been previously learned. The principle was first proposed by Amari in 1972 [7], and the network's structure, including its applications, was demonstrated by Hopfield in 1982 [6]. The potential applications of this technology have been explored, including the development of associative memory and noise cancellation algorithms. Furthermore, the efficacy of this approach in solving combinatorial optimization problems has been demonstrated.

Figure 4.13b shows the mechanism of state transitions, called the Amari-Hopfield model [7]. We consider n elements ($x_i(i = 1 \sim n)$), each taking two values. At time t, they are $x_i(t) = 0$, or $x_i(t) = 1$. Here, we assume a network in which each of the two elements x_i and x_j is connected to the other element, the connection coefficient w_{ij} is symmetric ($w_{ij} = w_{ji}$), and there is no connection to itself ($w_{ii} = 0$). Then the input u_i to the element x_i is $\sum_{j=1}^{n} w_{ij} \cdot x_j$

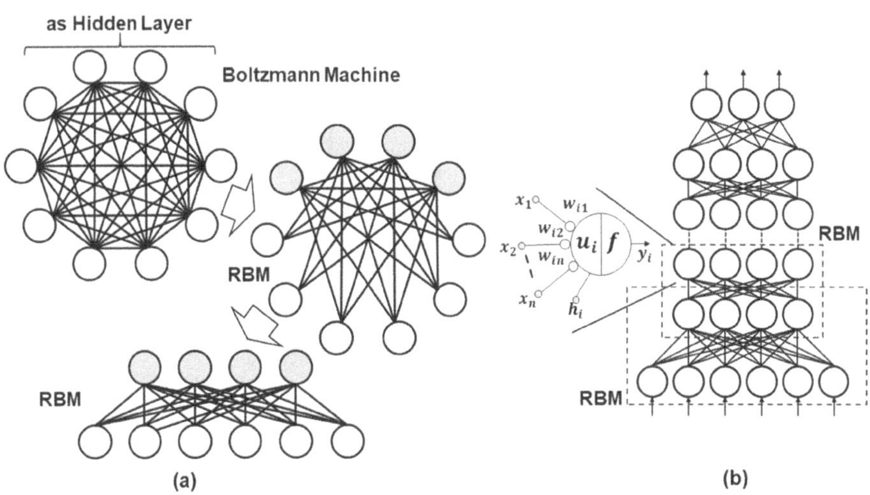

FIGURE 4.13
Hopfield networks and Boltzmann machines. (a) Restricted Boltzmann machine (RBM) and (b) deep belief network (DBN).

from the inputs from the other elements. Let the threshold here be h_i, and let x_i and u_i at time t be $x_i(t)$ and $u_i(t)$, then the same form as in equation (2.1) is obtained as;

$$u_i(t) = \sum_{j=1}^{n} w_{ij} \cdot x_j(t) - h_i(t) \tag{4.9}$$

Therefore, in terms of temporal behavior, $x_i(t+1)$ at time $t+1$ will behave as follows.

$$x_i(t+1) = \begin{cases} 1 \ u_i(t) > 0 \\ x_i(t)u_i(t) = 0 \\ 0 \ u_i(t) < 0 \end{cases} \tag{4.10}$$

This causes the system to vary in time in the direction of decreasing this energy when the following energy function is introduced.

$$E = -\frac{1}{2}\sum_{i=1}^{n}\sum_{j=1}^{n} w_{ij} \cdot x_i \cdot x_j + \sum_{i=1}^{n} h_i \cdot x_i \tag{4.11}$$

In other words, if we look at an element x_k, the first thing to look at is the following.

$$E = \left(\textit{terms where } i \neq k,\ j \neq k\right) - \frac{1}{2} x_k \sum_{j=1}^{n} w_{kj} \cdot x_j - \frac{1}{2} x_k \sum_{i=1}^{n} w_{ik} \cdot x_i + h_k \cdot x_k \tag{4.12}$$

Therefore, if we put $\Delta x_k = x_k(t+1) - x_k(t)$ here, the term $i \neq k,\ j \neq k$ will not contribute, as,

$$\Delta E_k = -\frac{1}{2}\left(\sum_{j=1}^{n} w_{kj} \cdot x_j + \sum_{i=1}^{n} w_{ik} \cdot x_i\right) \cdot \Delta x_k + h_k \cdot \Delta x_k$$

$$= -\frac{1}{2}\sum_{j=1}^{n}\left(w_{kj} + w_{jk}\right) \cdot x_j \cdot \Delta x_k + h_k \cdot \Delta x_k$$

Therefore,

$$\Delta E_k = -\left(\sum_{j=1}^{n} w_{kj} \cdot x_i(t) - h_k(t)\right) \cdot \Delta x_k = -u_k \cdot \Delta x_k \tag{4.13}$$

and from equation (4.10), ΔE_k is always negative. In other words, the energy is always decreasing. Here, since each neuron is coupled with all couplings in both directions, each coupling is assigned a weight, so that a change in the state of one neuron will affect all others connected to it.

However, the challenge arises from the fact that, given the transition's deterministic nature in a single direction, the energy post-transition does not attain a global minimum value; rather, it attains a local minimum value. If the value is locally small, the transition will not proceed any further. This leads to obtaining a solution that is not optimal for the given combinatorial optimization problem.

Therefore, the Boltzmann machine [8] was developed, in which the state transitions were made deterministic in the Hopfield network, but were developed into stochastic behavior. This approach enabled the execution of operations with the objective of minimizing the energy function. In the Boltzmann machine, the transitions of the above element x_i are not (3.10) but stochastic transitions. That is, when the energy difference in the transition of the element x_i between the next 1 and the next 0 is set to ΔE, this transition is assumed to be 1 with the following probability.

$$P(x_{i+1} = 1) = \frac{1}{1 + e^{-\Delta E/T}} \tag{4.14}$$

We introduce the concept (parameter) temperature T so that the larger (higher temperature) this parameter is, the more random this transition will be. At temperature 0, this is the same as the Hopfield model. You can also use $e^{-\Delta E/k_B T}$ for the exponential part of the function to match the format of statistical thermodynamics. Here, k_B is called Boltzmann constant. Note that this will be the same as the state transition with annealing in the Ising machine described in Chapter 9.

Historically, this Boltzmann machine developed into a restricted Boltzmann machine with no connections between the same layers, as shown in Figure 4.14, leading to a multilayer NN. In the restricted Boltzmann machine depicted in Figure 4.14a, in contrast to the mutually free connections observed in a Boltzmann machine, a restriction was implemented to confine the network to two layers: an input layer and a hidden layer. This restriction was introduced to enhance the convergence of learning. It became evident at this juncture that the hidden layer comprised features extracted from the data. This finding also facilitated comprehension of the function of the intermediate layer in multilayer NNs. The multilayered restricted Boltzmann machine shown in Figure 4.14b is also called a deep belief network (DBN). In a DBN, there are connections between layers, but there are no connections between units within each layer. Furthermore, by introducing directionality in the signal flow and adding an output layer, the structure becomes identical to that of a feedforward network. In a feedforward network, signals progress from the input layer to the output layer, and intermediate layers do not necessarily need to have a clear layer structure. However, in a DBN, the intermediate

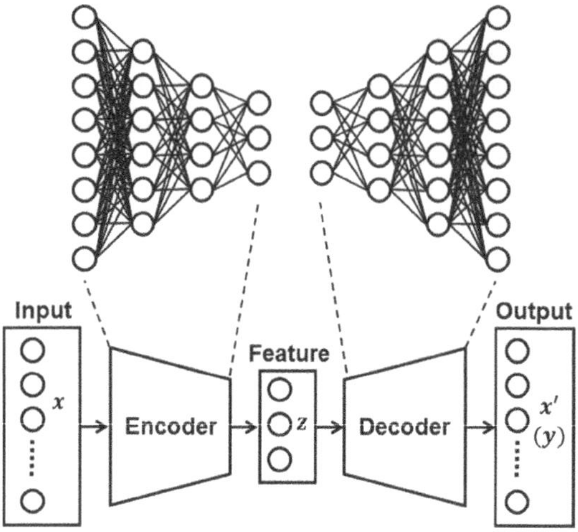

FIGURE 4.14
RBM and DBN.

layers are defined as having a layer structure. In this layer structure, each layer is understood to represent more abstract features of the input data as the layers progress. Furthermore, it has been demonstrated that by pretraining each layer and then fine-tuning the entire network, higher recognition rates can be achieved through multilayering. A challenge in LSI implementation is that, although the network is symmetric, there are interactions between all elements. This results in a large number of connections, making it difficult to implement the network as a circuit in its original form.

The Hopfield network and Boltzmann machine have the same coupling structure and system behavior as the Ising machine in the all-coupling case described in Chapters 6 and 7. The Ising machine does not like local solutions and seeks the minimum of the energy function, so it is further elaborated for this purpose.

4.4.4 Autoencoder

The autoencoder [9] consists of two parts: an encoder and a decoder, as shown in Figure 4.15. The encoder has an input (x) with n elements and an output (z) with fewer (k) than this. The decoder, which then takes this as input, uses this NN in reverse. This produces an output (x') with the same n elements as the first input, and the first input is used as the teacher data for this output. Suppose that the output (x') is the same as the original input (x). Then this output (z) will have less elements (dimensionality) than the input (x), but will have the original information compressed. If this encoder is multilayered, it can extract higher-dimensional features of the data and learn more abstract representations. This will be applied to natural language processing.

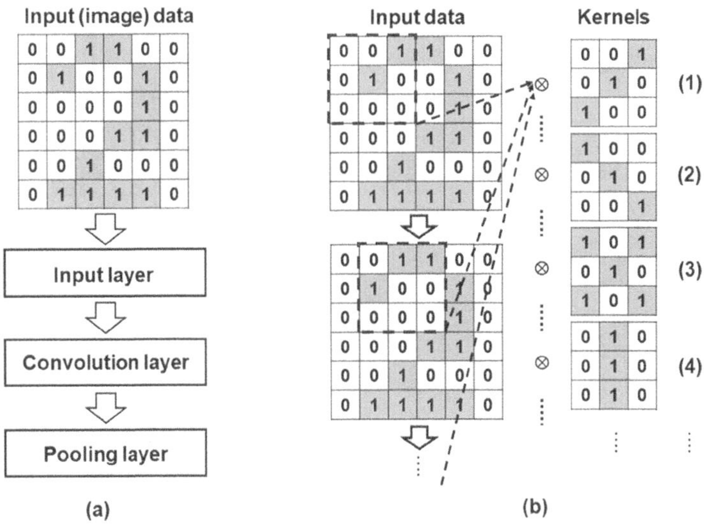

FIGURE 4.15
Variational autoencoder (VAE). (a) Convolutional network and (b) convolution processing.

4.4.5 Recurrent Neural Networks

When it is necessary to store and process time series data, which is sequential information or ordered data, conventional NNs are difficult to use because they process fixed-size vectors as input and generate fixed-size vectors as output. Recurrent neural networks (RNNs) [1] are a method of improving the performance of a network for subsequent inputs in a time series by using information that has already been signal-processed. As shown in a simplified form in Figure 4.16a, the network contains a portion that stores past information and internally loops through this information; the middle layer of the RNN has a path that conveys the output from the middle layer at one time to the middle layer at the next time. The intermediate layer at time t thereby receives input from the input layer at the same time t as well as input from the intermediate layer at the previous time t-1. Thus, the error back propagation method cannot be applied as it is, because the time lapse of the intermediate layer must be taken into account. Therefore, the BPTT (backpropagation through time) method shown in Figure 4.16b is an improved method for RNNs that can represent time lapse. This is a method in which the RNN network is expanded in the time direction through the intermediate layer outputs. It makes it possible to apply the error back propagation method by considering the RNN as one large NN, including the passage of time.

Recurrent NNs with long short term memory (LSTM) have also been developed, which can decide whether and how far back to use past information through learning. This has led to their use in natural language word processing, text analysis, machine translation, speech recognition, and image and

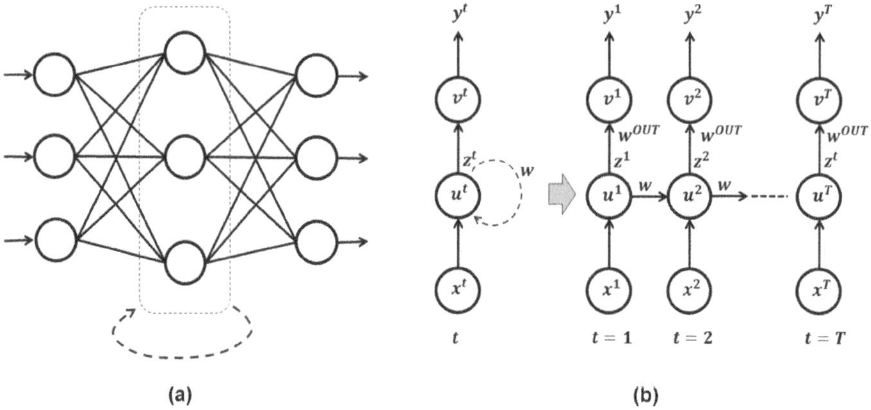

FIGURE 4.16
Recurrent neural networks (RNN). (a) RNN and (b) processing time-domain data using RNN.

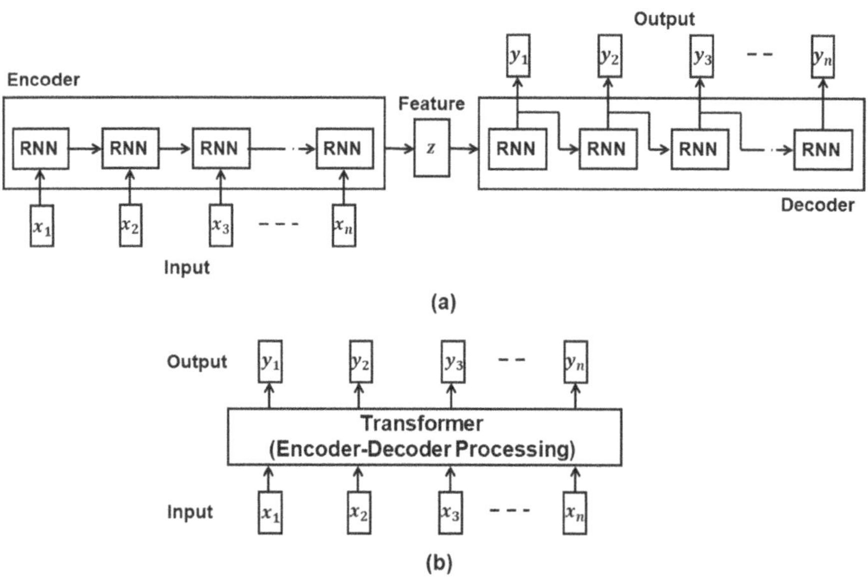

FIGURE 4.17
Processing of time-series data. (a) Sequential processing of time series data using RNN and (b) simultaneous processing of time-series data.

video analysis. RNNs are also often used in LSI applications, especially for computationally resource-intensive processing.

Figure 4.17a shows an example of a system using this RNN, particularly important for natural language processing. This is also a system using an encoder and a decoder. The inputs, x_1, x_2, x_3, ... x_n, are time series or series data, which are data given in this order. For example, it is the sequence of words in a single sentence from the beginning of the sentence. The encoder updates its internal state (hidden state) by reading these in sequence. This update is based on the previous internal state and the current input data. The last internal state information, z, can be regarded as a compressed representation of the entire input series. This is passed to the decoder. The decoder sets the final state$z$ received from the encoder as its initial state. When a start signal is given to it, the internal state changes, and by using that state as the next input, the serial data outputs, y_1, y_2, y_3,... y_n, can be obtained.

However, RNNs are no longer used much in large-scale neural networks: RNNs can handle time-evolving data, but they are slow and cannot be parallelized because LSTM is a sequential processing mechanism. In addition, the LSTM method has a limit to the length of time it can handle, and the model has become more complex as it tries to overcome this limitation. On the other hand, as a result of the development of hardware in the cloud that implements large-scale NNs, it is now possible to handle time-series data as is, and a configuration suitable for this has been devised. As shown in Figure 4.17b, this is a method in which the encoder and decoder are processed together, and the data is converted into a format that is easy to process in the encoder section, and the output of the encoder is processed in the decoder.

4.4.6 Transformers

When RNNs tried to handle information across long intervals, the information was not conveyed. Also, RNNs process information one by one according to the order of sentences, so to get the information for the next word, you have to wait until the previous word is finished, which makes parallel processing difficult and computation slow. To solve this problem, transformer [10] calculates the relationship of each word to all other words, as shown in Figure 4.18a, and processes based on this important information. For this purpose, time-series data such as natural language is first put into a format that can be processed in parallel by positional encoding, as shown in Figure 4.18b. Next, a function called attention is used. Attention mechanism is a method of processing information by concentrating on the important parts, allowing the encoding and decoding process to focus on specific parts. It also prevents information from fading when processing longer sentences or data, and allows the user to concentrate on important information. In addition, it became clear that it is possible to grasp more important information by identifying the relationships between parts in a series of sentences. The technology is called self-attention mechanism. This allows, for example, more natural sentence generation by taking into account the relevance of the previous and subsequent parts of the sentence. Furthermore, this is parallelized and called multihead. These technologies have advanced natural language processing, and in everyday applications, excellent progress has been made in machine translation, sentence generation, and question answering. In translation and other applications, an attention mechanism is processed between the source and target texts (cross-attention mechanism).

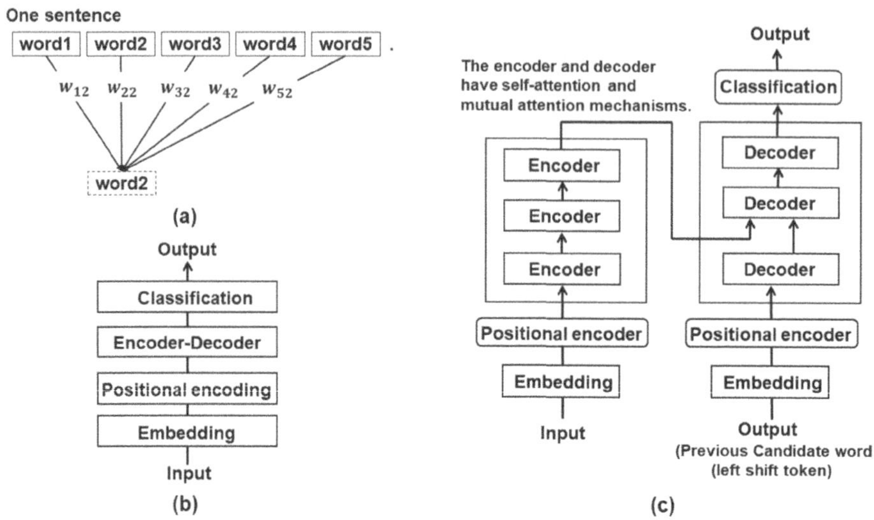

FIGURE 4.18
Transformer. (a) Attention, (b) Encode-decoder processing, and (c) transformer structure.

Such large-scale models will be processed in the cloud. On the other hand, the artificial intelligence processing integrated circuits described in this book are capable of fully demonstrating their advantages at the edge, so cooperative operation between the cloud and the edge will be important. In terms of low power consumption, low traffic, and security, the direction is to reduce processing on the cloud while using large-scale models, and to focus mainly on processing at the edge.

Now, neural networks have developed from learning from the brain's information processing mechanisms. RNNs are said to resemble the time series processing of the brain's neural circuits, and attention is said to be related to the brain's attention mechanism. However, it is challenging to achieve the energy efficiency of the brain's information processing with these, which are said to operate at around 10–20 watts. It is thought that extremely low-power integrated circuits will be required, but a solution is not yet in sight. The next chapter will deal with low power, but the main focus here will be on utilizing the characteristics of current neural networks.

4.5 Summary

In this chapter, the basic structure of artificial intelligence (AI) LSIs was described. The LSI of a neural network with inference function is realized by densely integrating many basic arithmetic circuits with the same function and efficiently connecting them. In a typical configuration, the layer structure of a neural network is represented by a two-dimensional array, which combines an input buffer (one-dimensional), weight data (array-like), arithmetic units such as multiplication and addition, and an activation function circuit that performs nonlinear processing. This enables high-speed and parallel inference processing. Next, we discussed the elements necessary to implement "learning" in neural network LSIs. Learning includes processes such as error back propagation and weight updating, but the processing is heavy, and at present, it is not reasonable to implement these in LSIs. Furthermore, typical neural networks such as feedforward networks, CNNs, RNNs, Hopfield networks, and Boltzmann machines were reviewed, and the progress of Transformer and others was also mentioned.

References

1. I. Goodfellow, Y. Bengio and A. Courville, *Deep Learning*, Cambridge: The MIT Press, 2016.
2. M. Glesner and W. Pochmuller, *Neurocomputers: An Overview of Neural Networks in VLSI*, London: Chapman & Hall, 1994.
3. L. Deng, "The MNIST Database of Handwritten Digit Images for Machine Learning Research [Best of the Web]," *IEEE Signal Processing Magazine*, vol. 29 , no. 6, pp. 141–142, 2012. https://doi.org/10.1109/MSP.2012.2211477
4. D. E. Rumelhart, G. E. Hinton, and R. J. Williams, "Learning Representations by Back-Propagating Errors," *Nature*, vol. 323, no. 6088, pp. 533–536, 1986. https://doi.org/10.1038/323533a0
5. D. Culler and J. P. Singh, *Parallel Computer Architecture: A Hardware/Software Approach*, 2nd ed. San Francisco, CA: Morgan Kaufmann, 1999.
6. Hopfield, J. J., "Neural Networks and Physical Systems with Emergent Collective Computational Abilities," *Proceedings of the National Academy of Sciences in USA*, vol. 79, pp. 2554–2558, 1982. https://doi.org/10.1073/pnas.79.8.2554
7. S.I. Amari, "Learning Patterns and Pattern Sequences by Self-Organizing Nets of Threshold Elements". https://doi.org/10.1109/T-C.1972.223477
8. D. H. Ackley, G. E. Hinton, T. J. Sejnowski, "A learning algorithm for boltzmann machines," Cognitive Science, vol. 9, no. 1, pp. 147–169, 1985. DOI: 10.1016/S0364-0213(85)80012-4
9. D. P. Kingma and M. Welling, "Auto-Encoding Variational Bayes," in *Proceedings of the 2nd International Conference on Learning Representations (ICLR)*, 2014. https://doi.org/10.48550/arXiv.1312.6114
10. A. Vaswani, N. Shazeer, N. Parmar, J. Uszkoreit, L. Jones, A. N. Gomez, Ł. Kaiser and I. Polosukhin, "Attention is All You Need," *Advances in Neural Information Processing Systems*, pp. 5998–6008, 2017. https://doi.org/10.48550/arXiv.1706.03762

5

Low-Power, High-Performance AI Chips Structure and Related Computing

Today, the Internet of Things (IoT) [1,2] is developing. This field is expected to expand into a large market, but sustainable development is important. In this IoT, it is necessary to utilize signal processing technology not only in the cloud but also on the "thing" side of the physical world, where sensors are installed, i.e., at the edge, to expand the sensor area and to perform ultra-low power information processing that is useful for high-speed feature extraction. Therefore, LSIs for artificial intelligence processing (edge AI) at the edge are effective, but they are required to have even lower power consumption and higher performance. This is because, although the power supply is limited at the edge, the data to be acquired will become increasingly complex from simple ones. This is also expected to smartly connect the physical world and the world of bits in terms of sustainability.

In this chapter, we will discuss low-power and high-performance artificial intelligence LSI chips structure. In addition, we mention here the increasing use of FPGAs in actual deployment, and we also discuss the progress of high-performance cloud computing, which can also be utilized to reduce power consumption, although only partially. In addition, we will also discuss the direction in which learning is also being attempted, although currently learning itself is done in advance using a system in the cloud and inference is performed using the weights obtained. Future technologies will also be discussed.

Note that when considering LSI integration, it should be noted that memory access accounts for the majority of the energy consumption of an integrated circuit system using multiple LSI chips. In general, the energy required for an LSI chip to access external memory is three orders of magnitude greater than the energy required for an additive operation on an LSI chip that also uses on-chip memory. This solution is not included in this chapter and is dealt with in Chapters 8–10.

5.1 Integrated Circuits in Edge AI

Edge AI is a type of edge computing described in Chapter 1. AI processing is performed at the edge and communication with the cloud is reduced as much as possible. Edge AI allows for a stronger connection between the physical

DOI: 10.1201/9781003599067-5

world and the cyber world with reduced communication: AI is mounted on IoT devices, sensors, and other terminals to perform in situ processing. This will achieve real-time performance, reduced communication costs, reduced power consumption, and enhanced security.

In general, there are many cases where a sufficient power supply system cannot be secured at the edge, and solar cells, for example, are used as a power source. Because of the importance of this point, the LSI of the neural network itself contributes greatly to low power consumption, but further reduction of power consumption is important. Installation space is also limited, so miniaturization is essential. On the other hand, the data to be acquired is becoming more complex, representing the physical world in greater detail. Only the essential and useful data should be extracted as data that can be processed.

Against this backdrop, the following is a summary of the major issues being considered for integrated circuits for edge AI.

1. Improvements in neural network structure: Sparse operation is being considered, which would take advantage of the redundancy of neural nets and change the structure of the network. This is a particularly significant development in inference. Also, in learning, this includes lowering power consumption by increasing the efficiency of data collection of the desired type and collecting more fine-grained data. To this end, a configuration is being considered that would include some of the learning functions and leave some parts of the system that can be changed in functionality with additional learning.
2. Lightweight neural networks: Quantization and low-precision methods that use the robustness of the network to reduce the precision of operations without reducing the precision of the results are being studied vigorously because of their significant power reduction effects. By limiting the application and purposely reducing the precision of data and weights, the amount of computation and memory usage on the LSI is also reduced, and energy consumption is reduced. Depending on the application, it is possible to reduce the number of bits in the weights while maintaining the accuracy of the results and improving the efficiency of the computation. Apart from this lower precision, the structure of the model can also be simplified to reduce the amount of computation and the number of parameters. For example, in convolution calculations, the computational complexity is reduced by separating the depth and spatial directions. Or, dimensionality reduction can be used to compress the weight data by converting it to the appropriate axes. Furthermore, in a technique called distillation, a computationally expensive model is used as a basis for creating a less computationally expensive model with the same functionality.
3. Hardware-level optimization: Optimizing architecture and circuit design to achieve high-speed and low-power operations for the edge.

> Although all of the numerous circuit-level low-power methods can contribute, based on an understanding of AI processing, attempts are being made to achieve higher performance and lower power consumption through slightly consolidated combinations of logical operations. Since the circuits are dedicated, it is an option to build all of them in a wired logic system. In addition, since this is the field of AI, which is evolving rapidly, and since verification at the level of actual LSI devices is required, FPGAs are often used.

In this chapter, we will look at sparse neural networks for structural optimization, low-precision neural networks for weight reduction and quantization, and FPGAs for hardware-level optimization that can demonstrate various ideas and lead directly to social implementations.

5.2 Sparse Neural Networks

There is a wide range of processing in edge AI. Neural networks can handle each of these applications by learning, but the values of the various weights also vary depending on the application. In this case, weights that do not affect the results such as discrimination accuracy will appear. At this time, this neural network is said to be sparsely coupled (sparse coupling) [3]. If there is no effect on accuracy, etc., the structure of the network itself will be strictly different, but it is possible to construct a neural network excluding these sparse connections. This method is also called branch pruning or branch cutting.

As we have seen, the two basic neural network equations are as follows:

$$u_i = \sum_{j=1}^{n} w_{ij} \cdot x_{ij} - h_i \tag{5.1}$$

$$y_i = f(u_i) \tag{5.2}$$

In a huge network, this n value is large. As an integrated circuit, the type of circuit itself is small, but it requires a circuit for calculating the sum of many products according to this value, wiring for connection, and driver circuits to drive this wiring.

Neural networks, on the other hand, can show complex functional relationships between inputs and outputs by modifying weights. In this sense, it is originally a reconfigurable technology. At this time, for example, if we construct a neural network that shows the relationship between the handwritten numeric data in the previous chapter and the numbers it indicates, there are often many weights whose value is close to 0. As is clear from equation (5.1),

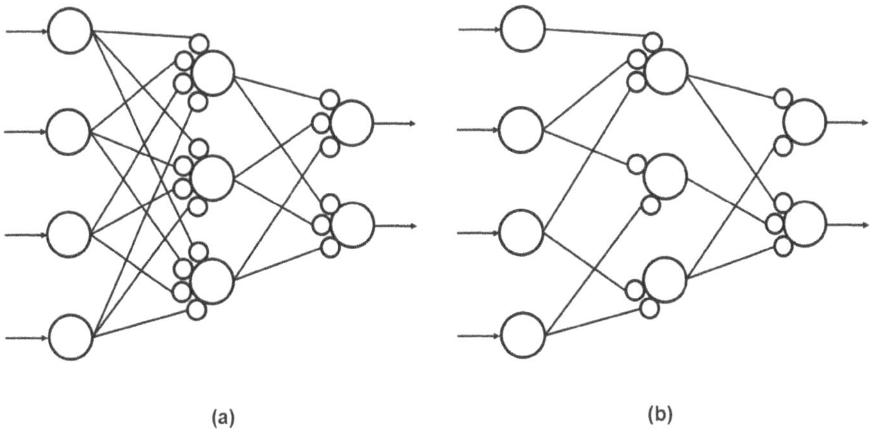

FIGURE 5.1
Scarification of neural networks. (a) Before learning and (b) After learning: Sparsification (pruning).

this term of weights will have a small contribution to the calculation of u_i. Therefore, if this neural network will only be used to identify handwritten characters after training, such terms can be omitted. How much to omit the smallest term is decided by the combination with the desired performance (recognition rate if it is a handwritten digit) while also performing regularization. Figure 5.1a shows the neural network before learning. If the terms with small contribution are omitted after learning, the network will have the structure shown in Figure 5.1b. In some cases, the neurons themselves are omitted. This means that, as an integrated circuit, the circuits and wiring related to this calculation are omitted. This allows for a smaller footprint and lower power consumption. The elimination of wiring is particularly effective. In some applications, it has been reported that even if as much as 80% of the weights are omitted compared to the learning process, the performance, such as discrimination during inference, is hardly degraded compared to the case where these weights are left in place.

This sparsification of the neural network is to incorporate only the necessary networks as dedicated to inference, which means that the size and structure of the network will be different during training and inference. There is also a concern that this will lead to over-specialization for exclusive use, even though the technology is reconfigurable. As a countermeasure, the circuit itself may be left in the same configuration. For example, the same network is used for inference and learning, but if the absolute value of the weights is below a certain value and close to zero, a function is added to stop the input operations. This would add additional functionality as an individual circuit, but would reduce the number of units operating during inference, resulting in lower power. It remains versatile and is not fixed to a location, time, or application, making it useful for data collection in a variety of locations.

A similar configuration is one in which weights that do not affect the results during learning are basically omitted, including the calculations, but in neural networks with multiple layers, for example, the function of updating weights for specific layers, i.e., part of the learning function, is retained. This allows the neural network to be used for inference in a specific application as an application to the IoT, but reduces the mounting area and power consumption while enabling changes according to the situation at the location where this neural network is placed. The same inference machine can be configured according to the individual situation in which it is used, even for a specific application. It can also accommodate changes in specifications during the development process.

5.3 Low-Bit-Precision Neural Networks

What is important in the consideration of neural network lightweighting is to strike a balance between the setting of the required calculation accuracy and the increase in circuit scale (increase in cost and power consumption) that accompanies the increase in calculation accuracy. If the calculation accuracy is low, in terms of neural network performance, for example, the image recognition rate will decrease. However, although it depends partly on the application, in fact, there are many reports that the actual recognition rate reduction relative to the calculation accuracy was kept low [4]. In the early days of neural network development, 32 bits were used, which was the logic operation and precision at that time. However, as a specific example, even if the precision was reduced from 32 bits to 8 bits in ResNet-50, a convolutional neural network with 50 layers in depth, the recognition accuracy of ImageNet, a data set for image recognition using ResNet-50, was reduced by only about 2% [5]. On the other hand, if this were implemented, a large reduction in power consumption and implementation size would be expected. This is because, to simplify, the scale of floating-point operations per image is about 1/4 of that for 8-bit operation compared to 32-bit operation. This example is an example of a large-scale neural network, but it has been reported that in a network where the required recognition rate can be obtained with an original 8-bit precision, even if the precision is reduced to 4 bits, the recognition rate will not decrease. In a specific application, there are several reports showing that even if the accuracy is reduced to 2 bits, the accuracy degradation in image recognition is 2%–3% compared to 32 bits. In addition, binarization [6], which expresses data as 1 bit, is also performed.

As shown in Figure 5.2a, development attempts continue to make bit accuracy as low as possible while maintaining such a recognition rate. Since inference is important at the edge, high accuracy is often used for training, and binarization is used for inference.

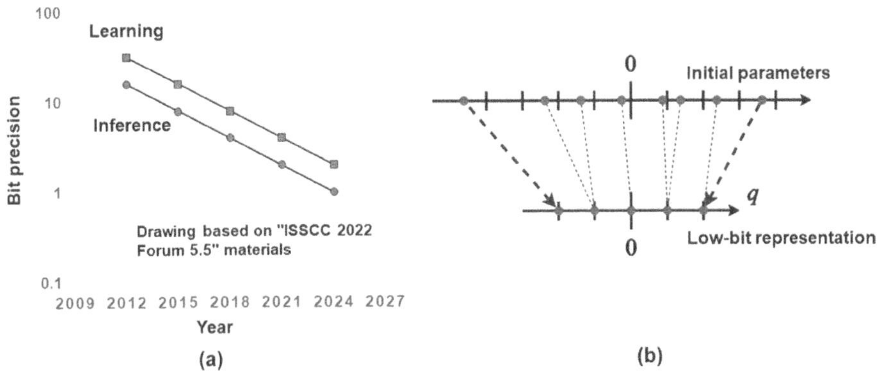

FIGURE 5.2
Low-bit neural network. (a) Progress of low-bit development with same accuracy and (b) Conversion to low-bit.

5.3.1 Conversion Method to Low Precision

As mentioned above, in low-precision learning, weights are often learned at 32 bits, for example, and the results are converted to lower bits for use in inference. Figure 5.2b shows the conversion as an image. In actual conversion, a threshold value, which is not shown in the figure, is set and compared with this value. At this time, logarithmization is also considered effective.

The threshold itself can also be obtained by learning to convert it to a lower bit that is suitable for a particular application and can achieve high accuracy in that application.

5.3.2 Binarization, XNOR Net, and Ternarization

Binarization, in which weights and bias values are represented by a single bit (0 or 1) in neural networks, is highly effective in reducing power and mounting area in LSIs. The challenge is that binarization limits the ability to express weights and the types of functions that can be expressed in the model, resulting in a lower performance, such as recognition rate, depending on the application. In addition, the discretization of the values of the weights also results in errors. However, reports of development to date indicate that the degradation in recognition rate is about 1%–5%. Considering applications at the edge, this decrease in recognition rate is often within an acceptable range.

If both the values and weights of this input and output are binary, the entire multiplication circuit can be replaced by XNOR. This is also called XNOR-NET [7]. The adder can also be replaced by a counting circuit. However, it is difficult to learn with only binary weights, and weights with higher precision bits are required. However, it has been reported that by introducing a scaling factorα during learning, multiplying with the whole sum of products, and updating it in the same way as the weights during the error back propagation method, when converted to binary weights for inference, it is possible to reproduce the change in values during learning using higher precision bits.

In ternarization [8,9], weights are represented by three values, −1, 0, and 1, to reduce the accuracy loss. This is because the expressive power of the weights can be increased by using three values: −1, 0, and 1. Also, the weights are effectively ternary while the input and output are binary. In an evaluation example using the dataset MNIST, the recognition rate was reported to be as high as 93%–96%, and the difference in accuracy from the case of real number weights was as small as 0.3%–0.6%.

From the point of view of reducing the implementation area and power consumption of LSI circuits, we describe in some detail binary neural networks (BNNs) and ternary neural networks (TNNs) and sparsification of TNNs.

Using the neural network in Figure 5.3 with one intermediate (hidden) layer as an example, we consider a discriminator whose input is a binary representation of each pixel in the dataset MNIST. Let us use the handwritten digit recognition MNIST as the dataset; the MNIST dataset consists of handwritten digit images from 0 to 9. The size of the image is $28 \times 28 = 784$, so the number of neurons in the input layer is 784, and the number of neurons in the output layer is 10 for 10-class classification.

As a BNN, we consider a configuration in which the weights and activation values are binary (−1,1), as shown in Figure 5.4. The binary weights and activation values can be determined by using thesgn function shown in equation (5.3).

$$\mathrm{sgn}(x) = \begin{cases} +1, & \text{if } x \geq 0 \\ -1, & \text{otherwise} \end{cases} \tag{5.3}$$

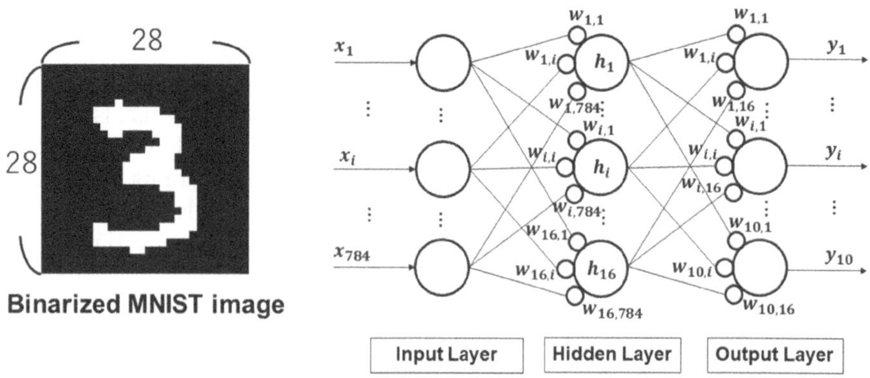

FIGURE 5.3
Three-layer neural network structure.

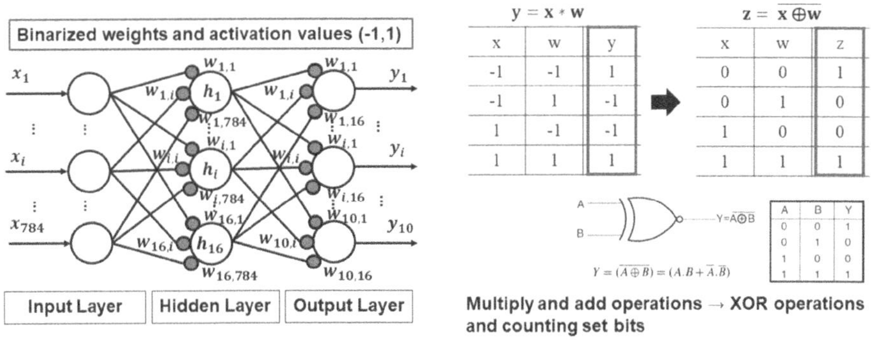

FIGURE 5.4
Binary neural network (BNN).

As shown in Figure 5.4, by treating "–1" as "0" in the circuit, multiplication between –1 and 1 can be replaced by XNOR, and the total input calculation can be replaced by counting up the standing bits. The total input calculation can be replaced by counting up the standing bits.

The TNN is then trivalent in terms of the possible values of the weights, as shown in Figure 5.5a. However, the output of each neuron by the activation function is assumed to be by the binary activation function. In this configuration, it is necessary to define appropriate thresholds when the weights are trivalued. If the real weights are W_i and the thresholds are Δ, the trivalent weights W_i^t can be determined as in equation (5.4).

$$W_i^t = \begin{cases} +1, & \text{if } W_i > \Delta \\ 0, & \text{if } |W_i| \leq \Delta \\ -1, & \text{if } W_i < -\Delta \end{cases} \tag{5.4}$$

TNN is expected to have higher recognition accuracy than BNN because of its wider range of weight expression compared to BNN.

Here, the weights are trivalued as (–1,0,1), and a weight with a value of 0 appears. As shown in Figure 5.5b, the function of weight 0 can be realized by not implementing the wiring for this weight 0. Therefore, in effect, the weights to be implemented on the circuit are (–1,1) and the activation value is also (–1,1), so binary multiplication can be replaced by XNOR as in BNN.

In this method, the weights are three-valued, which is more expressive than binary weights and is expected to improve recognition accuracy. Moreover, for weight 0, the function of weight 0 can be realized, and the circuit size can be reduced by not implementing the wiring of the part leading to it on the circuit. Examples of these effects are shown below. Table 5.1 shows an example of the recognition accuracy for MNIST test data verified with three patterns of weights: real, trivalent, and binary [9]. All are binary for input and

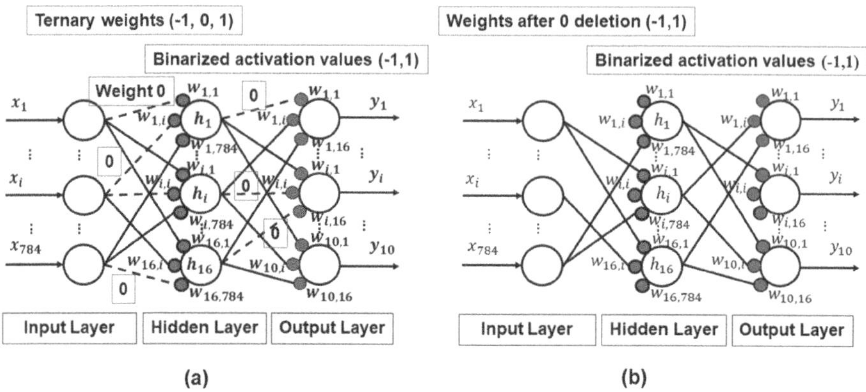

FIGURE 5.5
Ternary neural network (TNN). (a) Weights and outputs and (b) wiring.

TABLE 5.1
Comparison of Recognition Accuracy on Mnist Test Data Fpga Board Used (Xlinx,Vertex-7,Xc7vx485tffg1761-2)

	Possible Values for Weights		
	Real Number	Tertiary Value	Binary Value
Recognition accuracy (%)	91.46	89.97	86.27
Type of neural networks	Recognition accuracy (%)		LUT usage rate (%)
Binary neural networks	86.27		4.78
Tanari Spars XNOR-Net	89.97		1.11

activation values. In the case of the binary weights, the recognition accuracy decreased by about 5.2% compared to that of the real weights, while the trivalent weights were able to suppress the recognition accuracy decrease to about 1.5%. It can be considered that the trivalent weights are more expressive than the bi-valent weights, resulting in improved recognition accuracy.

5.4 Hardware Optimization

The relationship between AI chips and ordinary computer hardware is described here. As shown in Figure 5.6a, the AI chip is connected to the BUS as a device that speeds up the AI processing part of the hardware in an ordinary computer, which exchanges instructions and data through a transmission path (bus) between an arithmetic unit (CPU) and a memory device (memory).

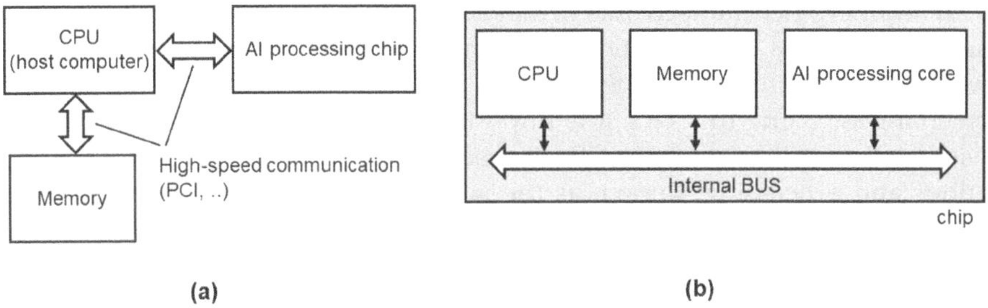

FIGURE 5.6
Hardware implementation form. (a) Discrete components and (b) SoC.

This allows the AI chip to perform a large number of sum-of-products operations and nonlinear operations at high speed, as we have learned.

This is because neural networks perform sophisticated signal processing by combining a large number of simple functions called neurons. On the other hand, signal processing in ordinary computers is aimed at general-purpose and high-speed processing, with memory and processors communicating data. This is not suitable for neural network processing. In other words, in this Neumann-type computer, the neural network is described as a program, and in order to execute it, instructions and data are exchanged through a transmission path (bus) between the arithmetic unit (CPU) and the memory device (memory). However, the speed of this transmission is the rate of performance and consumes a lot of power. To some extent, this can be addressed by building a small-capacity cache memory that can read and write at high speed into the arithmetic unit, and by setting it up in layers (1st cache/2nd cache/3rd cache) so that the same instructions and data do not have to be repeatedly retrieved from the memory device. However, in neural network processing, weight parameters are transferred from the storage device to the arithmetic unit, and a large number of sum-of-products operations are performed. At this time, since there are many parameters, it is necessary to perform memory accesses many times. This data transfer can consume more than three digits of power for the operation itself, which is a huge waste. Therefore, having this process performed by a dedicated chip is highly effective in increasing speed and lowering power consumption.

SoC is a technology that uses microfabrication techniques to put a system on a single chip, as shown in Figure 5.6b. In this case, AI processing is faster than in Figure 5.6a because the exchange between the arithmetic unit (CPU) and the memory unit (memory) is faster and less power-consuming. It should be noted, however, that AI processing is performed on a large amount of data. Therefore, if the data required for processing is in a different memory, even on the internal BUS, the performance is determined by the data transfer. To solve this problem, it is important to mount the content realized in the AI chip as the core of AI processing. The issue of the speed of this transmission is also discussed in Chapter 8.

In addition, neural networks achieve versatility by changing the weights, and the physical configuration of the hardware can be the same. In other words, programming is done spatially, and this structure does not need to be changed. Using this fact, it is important to optimize with a method that uses logic circuits combining physical elements and wiring to perform operations and processing, known as the wired logic control method, although some ingenuity is required in the handling of memory for parameters such as weights. In terms of computer resources, the wiring logic control method reduces the granularity of computation and enables highly parallel processing, thus making efficient use of computing resources. As a result, efficient use of resources is possible while maintaining high processing performance. The techniques described in Sections 4 and 5 of Chapter 3 are also important here as circuit-level low-power methods.

5.5 FPGA

As we have mentioned, a neural network was a signal processing in which simple functions called neurons are combined to perform sophisticated signal processing. Its operation has many parts that can be processed in parallel. Simply, if there are 256 neurons in one layer, 256 parallel operations are possible if each neuron is realized by an arithmetic unit. In this respect, field programmable gate arrays (FPGAs), which are equipped with a large number of arithmetic units, can easily be effective in improving the power-to-performance ratio. As development progresses and FPGAs become more readily available, it is easy to improve power consumption, processing speed, and element scale through configuration innovations. Also, good ideas can be verified quickly. Because optimization at the hardware level can be demonstrated quickly, it is attracting particular attention. While studying the high-level synthesis of logic circuits, called the logic circuit level or system level, this can also be used to study the circuit level by advancing to custom LSI design and taking advantage of the characteristics of transistors. In addition, FPGA products are expanding, and design tools are becoming more extensive and inexpensive. Therefore, FPGA [10,11] can be used to immediately verify the designer's ingenuity, contributing to the development of integrated circuits for artificial intelligence. In addition, it is said that GPUs used as coprocessors in servers can achieve power consumption one digit lower by replacing them with FPGAs.

However, at this time, knowledge of the hardware realization of neural networks is necessary. One of the advantages of recent FPGAs is that, once software is written, the tools provided with the FPGA can automatically generate hardware. However, in order to realize high-performance and low-power superior hardware, it is important to know the principles of FPGA and to incorporate the features of neural networks described so far. Here, we show a simple principle of FPGAs.

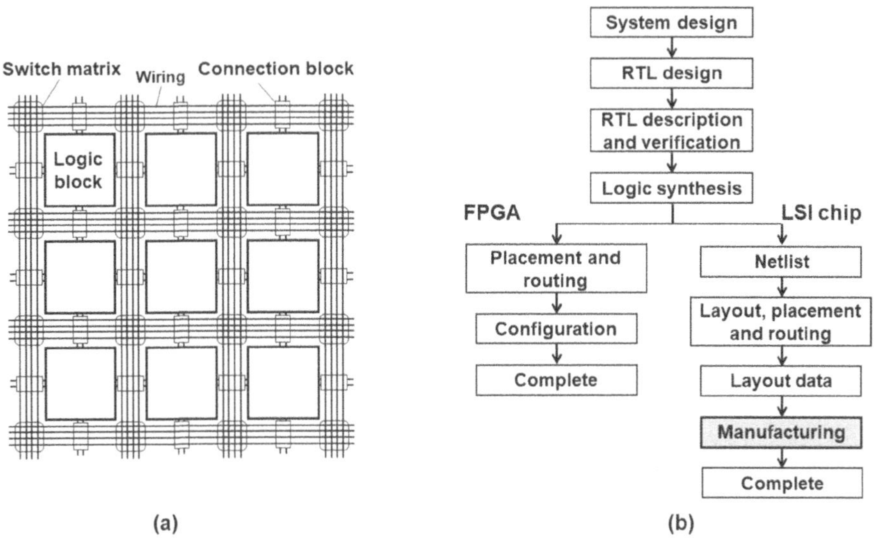

FIGURE 5.7
FPGA configuration and its design flow. (a) FPGA configuration and (b) design and implementation flow.

An FPGA is a programmable logic device. A typical FPGA consists of the Switch Matrix, Connection Block, Logic Block, and wiring elements as shown in Figure 5.7a. The Switch Matrix is the component that specifies signal paths and controls the paths within the FPGA. This Switch Matrix contains switches that connect input signals to output signals, and when the FPGA is programmed, specific connections are set up. This allows signals to be freely routed and controlled between different logic blocks. The connection block serves to connect the FPGA's logic blocks and wiring. The Logic Block, described below, is then the basic building block of the FPGA and is the unit for implementing programmable logic. Wiring is the wiring structure that connects the various elements in the FPGA. Wiring is used to transmit signals between logic blocks and usually consists of horizontal and vertical channels, which allow signals to be carried efficiently.

There are similarities and important differences in the design of an FPGA and a dedicated LSI chip, as shown in Figure 5.7b. First, the following steps are common: in system design, the architecture of the entire system is designed and necessary functions are defined; in RTL design, the design is done at the register transfer level (RTL) to abstract and describe the (digital) circuit operation. This RTL description is then used to verify that the design operates as intended. At this stage, the accuracy of the design is verified through simulation. Then, as logic synthesis, a logical circuit structure is generated based on the RTL description. This transforms the design into a form that can be implemented as hardware.

After this, the two differ.

In FPGAs, after Logic synthesis, placement and routing are the processes of placing logic elements inside the FPGA and routing them to connect them. This process is usually automated, but it should be noted that it depends on the characteristics of the FPGA. After this, the FPGA can be configured to the desired FPGA as a configuration. The design is now complete, and the FPGA is ready to perform its specific task.

On the other hand, in the design of a dedicated chip, after Logic synthesis, the first step is to generate a Netlist that contains the connection information of the logic circuit. This is important data that shows how each element of the circuit is connected. Then, the physical layout is designed, and each element is placed and routed. Here, it is also necessary to consider the constraints of the manufacturing process. This completed layout data contains the information necessary for manufacturing. This step is specific to dedicated chips and does not exist for FPGAs. This can take several months. Once the manufacturing is complete, the actual chip is finished.

The Logic Block, which makes the FPGA a programmable logic device, is shown in Figure 5.8. The basic unit consists of a LUT, a flip-flop, and the circuitry that connects them. In an integrated circuit, a look-up table (LUT) is a circuit that returns a predefined output for a specific input. The example in Figure 5.8 has four inputs (*A, B, C, D*) and produces one output (*Y*). The operation of this LUT is defined by the table in Figure 5.8a. This truth table shows what the output will be for each combination of inputs. For example, if the inputs are a particular combination (e.g., $A = 0, B = 0, C = 0, D = 0$), M1 is determined as the output Y based on the truth table. This M1 and other information is stored in a 16-bit SRAM and is selected via a selector. In this case, as shown in Figure 5.8a, the four inputs provide $2^4 = 16$ possible combinations,

A	B	C	D	Y
0	0	0	0	M1
0	0	0	1	M2
0	0	1	0	M3
0	0	1	1	M4
0	1	0	0	M5
0	1	0	1	M6
0	1	1	0	M7
0	1	1	1	M8
1	0	0	0	M9
1	0	0	1	M10
1	0	1	0	M11
1	0	1	1	M12
1	1	0	0	M13
1	1	0	1	M14
1	1	1	0	M15
1	1	1	1	M16

(a)

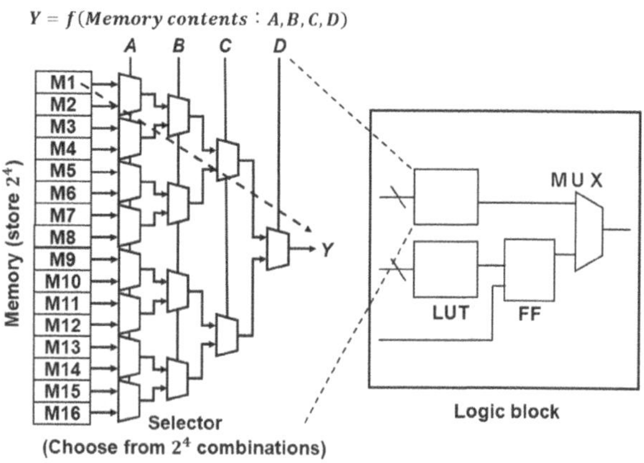

(b) (c)

FIGURE 5.8
Logical representation and implementation of FPGA. (a) Memory contents selecting table (b) LUT and (c) Logic block components.

and the Logic Block consists of a flip-flop (FF) and multiplexer (MUX) in addition to the LUT, as shown in Figure 5.8b and c.

The use of FPGAs facilitates circuit design and modification. For example, multiple modules can be designed and combined to build complex circuits. It is also easy to reuse circuits, and the same circuit can be used multiple times to achieve different circuit configurations. In addition, FPGAs can help optimize power consumption. For example, turning off unnecessary circuits can reduce the power consumption of a circuit. Also, by optimizing circuit elements and wiring networks, the number of circuits required can be minimized, thereby reducing power consumption.

Therefore, by designing the neurons that are the basis of neural network processing as modules and examining their connections, it is possible to examine them as a high-level synthesis of logic circuits. However, to repeat, FPGAs are generally equipped with a function that automatically optimizes the result of high-level synthesis from the viewpoint of the efficiency of using the FPGA's hardware. Over-reliance on this may not always reflect the designer's intentions correctly.

In addition, when processing large amounts of data, data transfer between the storage device that stores that data and the FPGA may become a bottleneck.

5.6 Quantum Computers

Quantum computers [12,13] are known devices using cryogenic temperatures, but there are also methods that can be used at room temperature using light or NMR. It can be assumed that these will be used at the edge. A quantum computer using NMR is already commercially available, albeit with a small number of qubits. If we follow convention and describe ordinary information processing as classical bit processing in a classical computer, a classical bit here is the basic unit of information quantity and is either 0 or 1. On the other hand, a qubit of a quantum bit is the quantum state $|\psi\rangle$ of the physical system that realizes the quantum computer that uses it, which is represented as $|\psi\rangle = \alpha|0\rangle + \beta|1\rangle$. The coefficients α and β are complex numbers, and $|\psi\rangle$ is a (quantum mechanical) superposition of the states $|0\rangle$, which is 0, and $|1\rangle$, which is 1. Here, the states $|0\rangle$ and $|1\rangle$ can also be represented as vectors $|0\rangle = \begin{pmatrix} 1 \\ 0 \end{pmatrix}$ and $|1\rangle = \begin{pmatrix} 1 \\ 0 \end{pmatrix}$, respectively. If we observe the quantum state $|\psi\rangle$, there for the first time we get the state $|0\rangle$ with probability $|\alpha|^2$ and the state $|1\rangle$ with probability $|\beta|^2$, which is $|\alpha|^2 + |\beta|^2 = 1$. We will not go into the physical implications of these here.

The use of these qubits enables parallel processing, which is essential for artificial intelligence processing. As shown in Figure 5.9a, the information

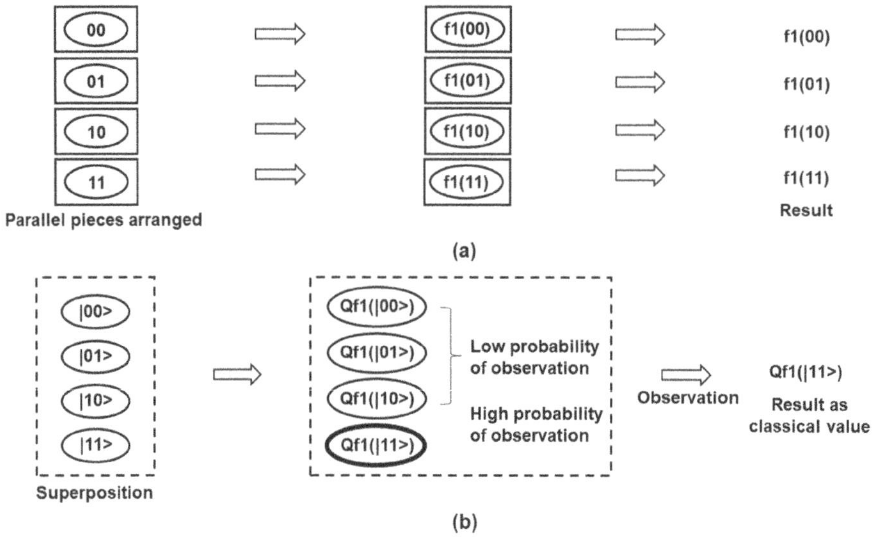

FIGURE 5.9
Parallel processing using quantum computers. (a) Classical computing and (b) quantum computing.

unit in a classical computer is called a bit, and the four bits b1, b2, b3, and b4 in the figure can only take 0 and 1, respectively. These are arranged side by side at the same time, which is parallelization in a classical computer. In other words, as shown in Figure 5.9a, 2 bits in classical computers can take the four values 00, 01, 10, and 11. To process these bits in a logic gate as inputs, they must be processed four times sequentially, or four logic gate circuits must be created, and each circuit must process one of them.

On the other hand, as shown in Figure 5.9b, the information unit of a quantum computer is usually called a qubit, which can take the superposition state of 0 and 1. Two bits can be represented as a superposition of $|00\rangle$, $|01\rangle$, $|10\rangle$, and $|11\rangle$. Thus, a single (quantum) logic gate can process them in parallel and simultaneously. By using this superposition state, multiple calculations can be performed at once in a single computation. However, in a quantum computer, the result of a calculation using this superposition state is determined by when it is read out (observed). Normally, it is not guaranteed that this is the result you want to obtain. Therefore, when performing calculations with a quantum computer, many calculations are performed in parallel using superposition states, and it is necessary to develop algorithms that take advantage of the properties of quantum systems so that the result obtained at the end is the desired result, and all other results disappear.

When an algorithm capable of such processing exists, the computational power of a quantum computer is extremely high. Historically, among these, an algorithm was discovered that could solve the discrete logarithm problem, which meant that prime factorization could be done at high speed. Since the difficulty of prime factorization was the basis of the widely used RSA encryption, quantum computers began to attract attention.

To apply this to neural network calculations, a hybrid system combining a classical computer and a quantum computer is currently being considered. As an example, the parameters of a quantum circuit (e.g., rotation angle with respect to a qubit) can be treated as equivalent to the weights of a neural network. These parameters change the output state of the quantum circuit, resulting in different candidate optimal solutions. Based on the measurement results obtained from the output of the quantum circuit, a classically defined loss function (error function) is evaluated. The parameters of the quantum circuit are adjusted by the classical optimization algorithm so that the value of this loss function is reduced. In this way, quantum circuits can search a large number of candidate solutions in parallel using superposition states, and thus are expected to optimize efficiently for a particular problem. However, high accuracy is required for constructing quantum circuits and obtaining the results of calculations. Error correction is necessary to deal with the effects of qubit errors and noise. Currently, error correction is being actively researched.

A few more basic facts about quantum computers are presented here. First, the quantum state $|\psi\rangle = \alpha|0\rangle + \beta|1\rangle$ can be expressed using Bloch sphere. As shown in Figure 5.10a, if we assume a point on a sphere of radius 1, and if we correspond the vector from the center of the sphere to the north pole to $|0\rangle$ and the vector to the south pole to $|1\rangle$, then $|\psi\rangle$, the superposition of these two vectors, is a vector pointing to any point on the sphere. Using this notation, the processing of a qubit corresponds to rotating a vector pointing to a point on the sphere around an arbitrary direction. Since a qubit can be any point on the sphere, it can be said to be an analog value, and if it contains noise, it is processed as it is in the arithmetic process. In the case of this sphere, it is not a point but a point that is blurred. It should also be pointed out that, due to the uncertainty principle, the sphere cannot exist as a true point in the first place. This is similar to the fact that in classical computers, analog computers cannot achieve infinite precision in terms of analog values, although the principle is completely different.

In a quantum computer, like the NAND gate in a classical computer, there is an element that serves as the basis for logical computation, and it performs an argumentation operation on the input and outputs an output. This is called a quantum gate, and Figure 5.10b shows this operation as U. As shown in this figure, the operation U is performed on one of the input qubits $|\psi\rangle = \alpha|0\rangle + \beta|1\rangle$, and the output $|\psi'\rangle = \alpha'|0\rangle + \beta'|1\rangle$ is produced. Here, the gate in a classical computer is a logical relationship between input and output, and also proceeds in one direction from input to output. However, in a quantum gate, the input qubit $|\psi\rangle$ represents a certain quantum state, which is changed to another quantum state as a result of the operation U, meaning that this becomes the output $|\psi'\rangle$. In general, the reverse process

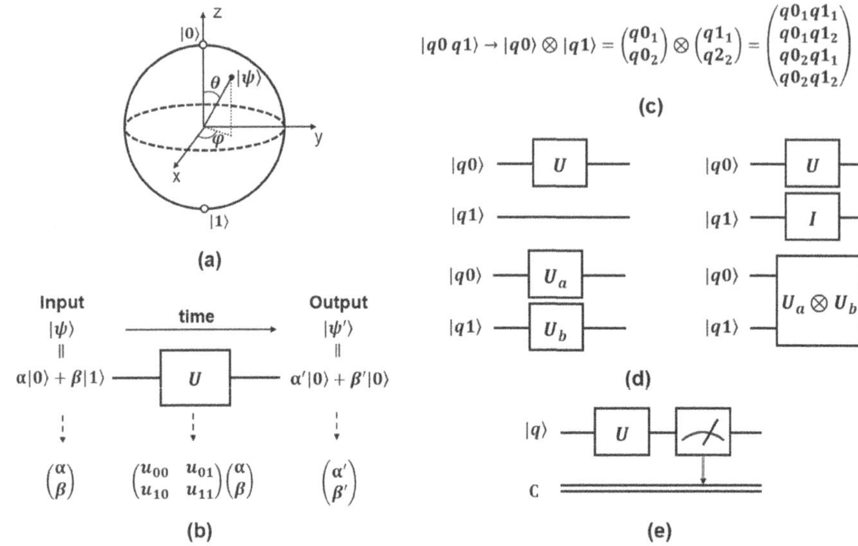

FIGURE 5.10
Basic quantum gate operation. (a) Bloch sphere, (b) quantum gate, (c) 2 q-bits, (d) 2 q-bits processing, and (e) Measurement.

also exists. In the representation of the Bloch sphere in Figure 5.10a, this operation U would be an operation where a vector pointing to one point on the sphere (input) is transferred to a vector pointing to another point on the same sphere of radius 1 (output). α, $\beta, \alpha', \beta'$are complex numbers, so this is a unitary transformation. The quantum gate$U$ is this transformation, which in this case can be represented by the 2-by-2 matrix shown in Figure 5.10b. A quantum gate is such a transformation on the input.

Next, we discuss multiple qubits. As shown in Figure 5.10c, for example, two qubits are represented as the tensor product of two qubits and their superposition as a state. Therefore, as shown on the right side of Figúre 5.10c, it is a vector with four rows and one column. This means that we have four of these bases. In other words, while $|0\rangle$ and $|1\rangle$ were the bases for the 1-qubit, the 4 bases for the 2-qubit are $|00\rangle, |01\rangle, |10\rangle, |11\rangle$, based on the tensor product of these two bases. Since these 2 qubits are superposition states, parallelism appears as already mentioned in Figure 5.9b. n qubits are the tensor product of n two-qubits, which means that they have 2^n bases. Quantum computation for more than 2 qubits is shown in Figure 5.10d. 2 Using the qubits as an example, suppose that the operation U is applied to only one of the qubits, while the identity transformation operation I is applied to the other qubit. As a matrix, it is a 2-by-2 unit matrix. Also, since the input 2 qubits was the tensor product of two vectors, it is equivalent to the operation $U \otimes I$ being performed on it. This is generalized under Figure 5.10c.

In a quantum computer, the result of a calculation using a superposition state is determined when it is read out (observed), so the operation of observation (readout) is important. It is only after this operation that the result can be obtained, i.e., retrieved as a classical bit. For this purpose, the operation of the symbol in Figure 5.10d is determined, and it is shown that the bit is read out as a classical bit at that time.

In future artificial intelligence applications, quantum circuits are expected to be used to implement algorithms such as quantum machine learning and quantum reinforcement learning. These algorithms are expected to use quantum circuits, as shown in Figure 5.10, to solve problems that cannot be solved by classical computers. Additionally, circuit designs employing quantum circuits and their operational outcomes are accessible to the general public via the cloud, provided by developers and research institutions. In other words, quantum computers are already available for practical use.

5.7 Brain Computers with Higher Performance and Learning Capabilities

Because of the wide variety of processing in edge AI, prior learning alone may not be sufficient for optimization, depending on the field being handled. This will affect not only computational performance (e.g., accuracy) but also power consumption. However, learning requires large computational resources. For this reason, it has been reported that only the layer before the output layer is equipped with a learning function to adjust the weights.

In addition, equipment placed in a natural environment is difficult to maintain on a regular basis once it is installed. On the other hand, conditions and environments change. Therefore, it is also intended to compensate for this by learning, and there are examples where a learning function is installed in cases where this analysis is relatively easy. For example, in the application of detecting abnormal vibration of a motor to detect signs of failure, a small neural network can be used to handle the problem, and real-time learning is said to be possible, and a system that starts additional learning autonomously in response to changes in the environment is being considered. In other examples of anomaly detection, instead of learning normal data, the system learns identifiers (features) for detecting anomalous data; the identifiers are learned by generating anomalous data in the LSI. For this purpose, abnormal data is generated by converting normal data to abnormal values, which are then used for learning.

These innovations enable learning on LSIs that can be used at the edge. This is necessary to achieve high accuracy and low energy consumption in real-time applications at the edge. This would also be a very effective technique from a privacy and security perspective. Furthermore, AI chips using the memory elements being described in Chapter 8 through Chapter 10 can be miniaturized and are being developed with high expectations for learning functions at the edge.

The trend toward higher performance is mentioned here as an example of the development concept. To run a neural network on an LSI is a mass of sum-of-products operations through matrix representation. Therefore, it is a matter of how to perform this at high speed and low power.

At this point, the structure of the data that is the target of the sum-of-products operation is important, and the solution is to build an arithmetic circuit suitable for this structure. The structure of the data is whether the sequence is one-dimensional, two-dimensional, or three-dimensional. Mathematically, a higher dimension can be considered. Also, when performing operations on two sets of data, the structure of these two sets of data is not necessarily the same. Therefore, it is important to create an arithmetic module that specializes in the most frequently used operation for the two data in the targeted artificial intelligence process and performs it at high speed and low power. These operations are then combined with the optimal smaller-sized arithmetic modules in the integrated circuit process used here to achieve the necessary functionality, based on the required precision. Note that this type of computation is also essential in video processing and was developed in integrated circuits for games even before the development of artificial intelligence processing became active. In the calculation of neural networks, the integrated circuit that efficiently performs matrix operations, which are operations between chunks of this data, is prepared after making the processing as parallel as possible at the higher levels in the program. Tensor Core [14] is known as an example. Tensor is the name for the entire matrix representation including higher dimensions as described in the previous chapter.

As described in Chapter 1, artificial intelligence processing and neural networks were originally separate technical fields, but with the advent of deep learning, they are now being developed as one. On the other hand, as will be described briefly in Chapter 11, there is the field of brain-based computers, which is a brain-based signal processing that aims to reproduce the brain itself or to process signals utilizing brain functions, in the development of artificial intelligence technology that is slightly different from but related to general artificial intelligence LSI. The development of integrated circuits that aim to model the neural circuits of the human brain is also underway. The main functions are the same as those of conventional neural networks, but the neural network called TrueNorth employs spiking neural networks [15] and makes extensive use of asynchronous operation. This allows for a hardware implementation of the neural circuitry, with multiple neurons and synapses on an integrated circuit, and low energy and high-speed parallel operations.

As shown in Figure 5.11, the spiking neural network receives a sequence of spikes as input and performs processing based on the timing information of the spikes, thereby achieving high-speed operations. The basic operation is

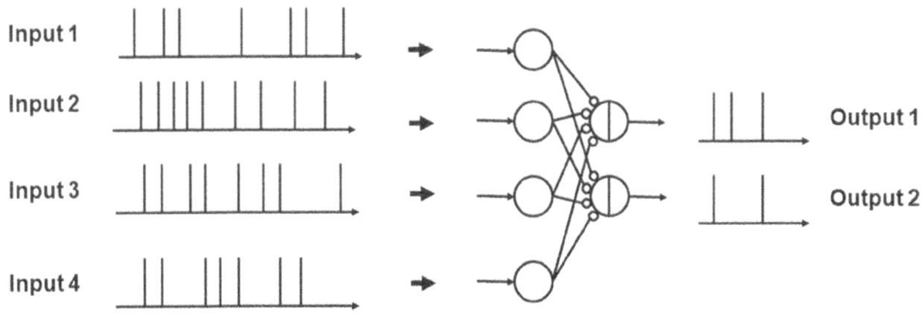

FIGURE 5.11
Spiking neural network.

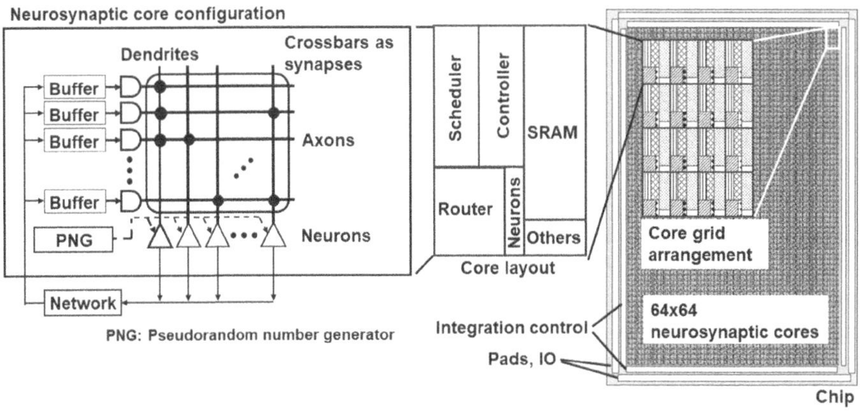

FIGURE 5.12
TrueNorth configuration.

to convert the input signal into a time-series binary signal (spike sequence) for processing. This processing can be done using only additive operations, and additive operations consume less power than multiplicative operations. Therefore, it is considered to be close to the behavior of the brain and to have low power consumption. In addition, the range of expression is wide, including not only the temporal density of a single input spike train, but also the relationship between multiple input spike trains. Furthermore, spike trains can be operated asynchronously, so to speak, when they are input, thus reducing unnecessary power consumption. From this point of view, it is said that low power consumption can also be achieved.

In TrueNorth [16], neural networks are implemented with the structure shown in Figure 5.12. In this example, each chip has a core of circuits that reproduces 256 neurons and 262,144 synapses, which is 64 x 64, or 4,096 neurons. The power consumption is reported to be 70 mW. The report further states that 16 of these chips are arranged on a single board to achieve a system with approximately 16 million neurons and 4.1 billion synapses.

This is said to be comparable to the number of neurons in the brain of an amphibian. While the entire system is operated at an extremely low clock speed, asynchronous operation here is event-driven with spike-sequence input. This achieves low power consumption and high performance.

The results of deep learning have led to significant advances in a variety of application areas, including automated driving, medical diagnostics, natural language processing, and image generation. Meanwhile, the development of brain-based signal processing and brain-based computers also aims to realize more advanced information processing and brain-computer interfaces by understanding the mechanisms and functions of the brain and mimicking the brain's information processing and learning.

5.8 Summary

The development of lower-power and higher-performance artificial intelligence LSIs can be approached from several directions. In the structure of neural networks, sparsification (pruning) is underway, which reduces the weight of the model. Additionally, the employment of neural networks with low-precision bits is of significance. The processes of binarization and ternarization are also being utilized. Hardware optimization encompasses the implementation of a wired logic control scheme, and ongoing development is underway that utilizes field-programmable gate arrays (FPGAs). The prominence of FPGAs stems from their capacity to expeditiously demonstrate hardware-level optimization. In the realm of high-performance computing, spiking neural networks are demonstrating significant advancements. The incorporation of learning functions in edge devices and the use of quantum computers are among the future trends that are anticipated.

References

1. H. Chaouchi, *The Internet of Things: Connecting Objects to the Web*, Hoboken, NJ: Wiley, 2010.
2. R. Buyya and A. V. Dastjerdi, *Internet of Things: Principles and Paradigms*, San Francisco, CA: Morgan Kaufmann, 2016.
3. S. Han, J. Pool, J. Tran and W. Dally, "Learning Both Weights and Connections for Efficient Neural Networks," *Advances in Neural Information Processing Systems*, vol. 28, 2015, pp. 1135–1143. DOI: 10.48550/arXiv.1506.02626
4. D. Miyashita, E. H. Lee, B. Murmann, "Convolutional Neural Networks Using Logarithmic Data Representation," arXiv:1603.01025, 2016.

5. B. Jacob, S. Kligys, B. Chen, M. Zhu, M. Tang, A. Howard, H. Adam, D. Kalenichenko, "Quantization and Training of Neural Networks for Efficient Integer-Arithmetic-Only Inference," *Proceedings of the IEEE Conference on Computer Vision and Pattern Recognition (CVPR)*, Salt Lake City, UT, 2018, pp. 2704–2713. https://doi.org/10.1109/CVPR.2018.00286
6. M. Courbariaux, I. Hubara, D. Soudry, R. E. Yaniv and Y. Bengio, "Binarized Neural Networks, Training Deep Neural Networks with Weights and Activations Constrained to +1 or –1," arXiv:1602.02830, 2016.
7. M. Rastegari, V. Ordonez, J. Redmon and A. Farhadi, "XNOR-Net: ImageNet Classification Using Binary Convolutional Neural Networks," arXiv:1603.05279, 2016.
8. Y. Yoshida, R. Oiwa and T. Kawahara, "Ternary Sparse XNOR-Net for FPGA Implementation," *7th International Symposium on Next Generation Electronics (ISNE)*, Taipei, 2018. https://doi.org/10.1109/ISNE.2018.8394728
9. T. Megumi and T. Kawahara, "Circuit Optimization of Ternary Sparse Neural Net," *IEEE 22nd World Symposium on Applied Machine Intelligence and Informatics (SAMI)*, Stará Lesná, Slovakia, 2024, pp. 53–58. https://doi.org/10.1109/SAMI60510.2024.10432809
10. R.l Merrick, *Getting Started with FPGAs*, San Francisco, CA, USA: No Starch Press, 2023.
11. U. Meyer-Baese, *Embedded Microprocessor System Design using FPGAs*, New York: Springer International Publishing, 2021.
12. N. S. Yanofsky and M. A. Manucci, *Quantum Computing for Computer Scientists,* Cambridge: Cambridge University Press, 2008.
13. M. A. Nielsen and I. L. Chuang, *Quantum Computation and Quantum Information,* 10th Anniversary ed. Cambridge: Cambridge University Press, 2010.
14. S. Markidis, S. W.-D. Chien, E. Laure, I. Bo Peng, J. S. Vetter, "NVIDIA Tensor Core Programmability, Performance & Precision," *IEEE Transactions on Parallel and Distributed Systems,* vol. 29, no. 8, pp. 1–12, 2018. https://doi.org/10.1109/IPDPSW.2018.00091
15. J. D. Nunes, M. Carvalho, D. Carneiro and J. S. Cardoso, "Spiking Neural Networks: A Survey," *IEEE Access,* vol. 10, pp. 60738–60764, 2022. https://doi.org/10.1109/ACCESS.2022.3179968
16. F. Akopyan, J. Sawada, A. Cassidy, R. Alvarez-Icaza, J. Arthur, and P. Merolla, "TrueNorth: Design and Tool Flow of a 65 mW 1 Million Neuron Programmable Neurosynaptic Chip," *IEEE Transactions on Neural Networks and Learning Systems,* vol. 23, no. 10, pp. 1459–1471, 2012. https://doi.org/10.1109/TCAD.2015.2474396

6

Ising Machines and Combinatorial Optimization Problem

Artificial intelligence (AI) chips are generally LSI versions of neural networks, as described in Chapter 2, but in parallel with this development, there is also progress toward other computational principles to replace traditional Neumann-type computing. Examples include quantum computing, neuromorphic computing, DNA computing, reservoir computing, and evolutionary computing. Within this technological trend, an important LSI technology has emerged not in the area of artificial intelligence or neural networks but historically in the context of quantum computer development. This is the Ising machine, which is the subject of this and the next chapter. However, even further back in history, this was actually developed as an early neural network, and the principle is the same as the Hopfield network and Boltzmann machine described in Chapter 4.

It is a dedicated computing method for solving combinatorial optimization problems [1].

6.1 Combinatorial Optimization Problems

Combinatorial optimization problems are problems for optimally allocating limited resources under given conditions and often appear in the real world. Here, "limited resources" and "given conditions" represent the objectives and strategies of the activity for which the problem is to be solved, and therefore, the optimization results and other information are highly confidential. In other words, we do not want to disclose the data of the conditions and results to the outside world. This also overlaps with the problem to be solved by artificial intelligence as described in Chapter 1. As Figure 6.1a shows an example, delivery and routing are the problems of determining the optimal route for delivering goods from a distribution center to a customer. In manufacturing, it is also a problem of determining which machine will be used to produce each product on the production line. The assignment is made in such a way as to maximize overall production efficiency while taking into account the capacity and production costs of each machine. It also includes the problem of optimizing the schedule of each worker. This will be the problem of

DOI: 10.1201/9781003599067-6

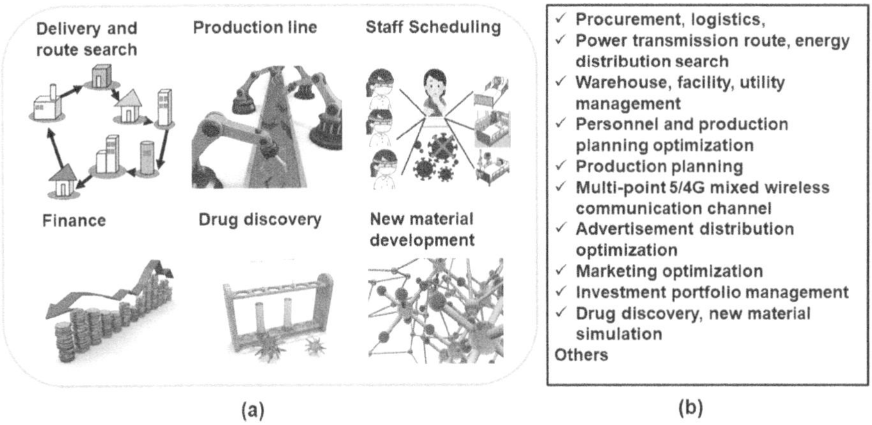

FIGURE 6.1
Familiar combinatorial optimization problems. (a) Combinational optimization problem (COP) and (b) Example of COP.

efficiently allocating the entire workforce while taking into account work hours, vacations, and skills. The same applies to logistics efficiency and cost reduction, including logistics management in warehouses. In finance, it is believed that risk management and optimization of investment strategies can be achieved in high-speed transaction processing. Furthermore, in drug discovery and materials development, it will be possible to examine a vast number of combinations. Figure 6.1b shows a number of other examples. These are manifested in many aspects, from daily life to business.

However, solving such combination optimization problems using conventional von Neumann-type computing by brute force (nonintelligent) search of all cases is impractical. For example, as the number of delivery destinations increases, the number of cases in which it is necessary to determine whether a delivery route is optimal increases exponentially, and the time (computational cost) required for this becomes impractical. In other words, if the number of delivery destinations is N, the number of cases is approximately N factorial. For four delivery destinations, the number of cases is only $4!=24$, but for eight delivery destinations, it becomes $8!=40{,}320$, and for 16 delivery destinations, it becomes 2×10^{13}, and for 24 delivery destinations, it becomes 6×10^{23}. In the case of 24 delivery destinations, even if a 1GHz processor were used to check one case per clock cycle, it would take over 10^{14} s (over 3 million years) to check all cases.

Of course, since this is an important area, there are dedicated algorithms that are not brute force. However, as shown in Table 6.1, this is diverse and would have to be addressed on an individual basis for each problem. Explanations on these methods are not given here, but the content of each one would be such that a specialized book exists on each one.

TABLE 6.1
Combinatorial Optimization Problem and Solution Method

Problem Name		Solution Method
Maximum cut problem	The problem of dividing the vertices of a graph into two subsets and maximizing the sum of the weights of the edges that are cut	Approximation solutions with guaranteed performance
Traveling salesman problem	The problem of visiting every given city once and finding the shortest route back to the original city	
Vertex cover problem	The problem of finding the smallest set of vertices needed to cover all edges of a graph	
Knapsack problem	The problem of choosing items that maximize the total value within a limited capacity	
Resource allocation problem	The problem of optimally allocating limited resources and minimizing costs	Greedy method
Minimum spanning tree problem	The problem of finding a tree that connects all vertices and minimizes the sum of the edge weights	
Minimum cost elastic matching problem	The problem of achieving an incomplete matching at the minimum cost	Dynamic programming
Shortest path problem	The problem of finding the shortest path between two vertices	Bellman–Ford algorithm Floyd–Warshall Algorithm
Maximum flow problem	The problem of considering flow constraints to maximize the flow rate in a network	Augmenting path method
Minimum cost flow problem	The problem of finding a way to flow through a network at the minimum cost	Negative cycle elimination algorithm
Maximum matching problem	The problem of finding the maximum matching in a graph	

On the other hand, this combinatorial optimization problem in general is isomorphic to the energy equation that represents a model called the fully coupled type, which is a development of the magnetic model known as Ising model [2] when the difference between the optimal solution and the trial time is defined as an error function. The energy equation can be used to determine the behavior of the Ising model, and this can be applied to speed up calculations that take advantage of this model. In this chapter, we will learn how to construct this Ising machine in an integrated circuit. If the spins are fully coupled, this is the same type as the Hopfield model described in Chapter 4, and what we call a fully coupled Ising machine here is the same as a Boltzmann machine in a broad sense. Therefore, as an integrated circuit, it can be considered in the category of LSI neural networks.

In this section, we will show how to solve the problem using the fully coupled Ising machine from a model of a magnetic material.

6.2 Ising Model

The Ising model is a statistical mechanical model that describes the properties of magnetic materials. In the Ising model, two orientations of spins are placed on a lattice point, and an interaction (coupling) is placed between the spins. The spins update their states according to the interaction between the spins and an externally given magnetic field. Eventually, the states of the spins converge so that the Ising model has the lowest energy state.

In fact, the combination of spin orientations that results in this minimum energy state is the solution that minimizes the error function in the combinatorial optimization problem. In this transition to the minimum state, therefore, it is not a brute force search of all states. There is a guideline for reducing energy. This can be accomplished with very little computational effort. In other words, this behavior mimics the behavior of the magnetic material, allowing for fast calculations.

However, it should be noted that this is an approximate solution, which should also be taken into account when considering LSI implementation. In other words, the accuracy of the calculation should be considered in the approximate solution, not in the exact solution.

6.2.1 Ferromagnetic Materials and Ising Models

Let us first look at the relationship between ferromagnetics and the Ising model. In the next section, we will look at the adjacent or sparse coupled and fully coupled types of spin coupling, which will be developed into a computational process.

The spins of electrons in a ferromagnetic body are collectively ordered and aligned in a small region. This region is called a magnetic domain, and each has spins aligned in a certain direction. The orientation of the spins can be thought of as corresponding to the S-pole to N-pole orientation in a bar magnet. A ferromagnetic material is one in which the spins are aligned even between each of the domains at a specific temperature range, and spontaneous magnetization develops. In this spontaneous magnetization state, magnetization occurs in the material itself even without a magnetic field.

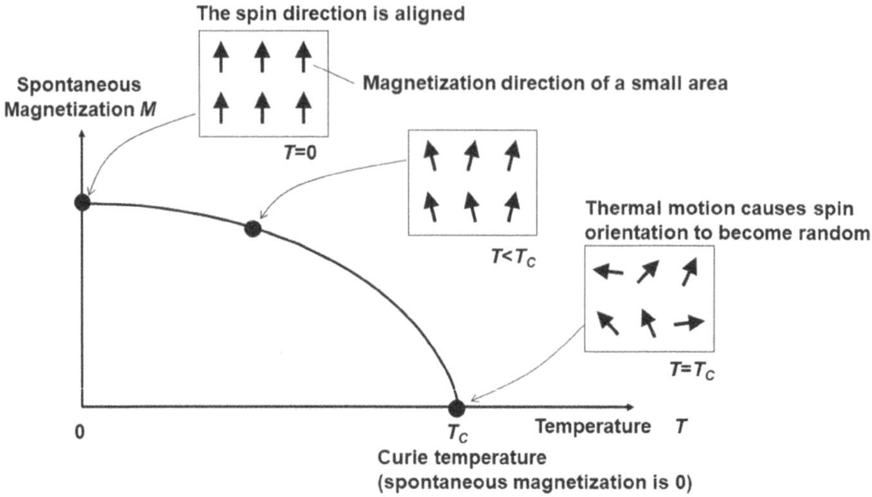

FIGURE 6.2
Relationship between magnetization and temperature in ferromagnetic material.

Figure 6.2 shows the relationship between spontaneous magnetization (M) and temperature (T). As shown in the figure, spontaneous magnetization decreases as temperature is increased, and at temperatures higher than the Curie temperature (T_c), spontaneous magnetization disappears. For $T < T_c$, spontaneous magnetization exists. Also here, the magnetization can be controlled by an external magnetic field. At absolute zero ($T = 0$), the spontaneous magnetization is maximal. The spins are perfectly aligned, and the maximum overall magnetization is achieved. This state is called the ground state and is the most stable. This mechanism is used to seek solutions to optimization problems. The most stable state corresponds to the solution of the optimization problem.

Figure 6.3 shows the ferromagnetic Ising model in two dimensions. This model was developed in the field of statistical mechanics to explain the fact that the ferromagnet shown in Figure 6.2 has spontaneous magnetization. Consider a two-dimensional lattice structure. The smallest unit of magnetization, the spin, is located at the lattice points, and there is an interaction between the nearest neighboring spins, and the orientation of the spins is also determined by the external magnetic field. This model describes the mechanism by which ferromagnetic materials are magnetized. In this model, the absence of magnetization in a ferromagnetic body at high temperatures corresponds to the fact that the spins are oriented in their own direction, as shown in (a). At low temperatures, as shown in (b), the spins are aligned in one direction, and thus magnetization appears as a whole.

To describe this phenomenon of aligning spin orientations, Figure 6.4 shows what happens in two spins. There are two spins (elements) σ_i and σ_j

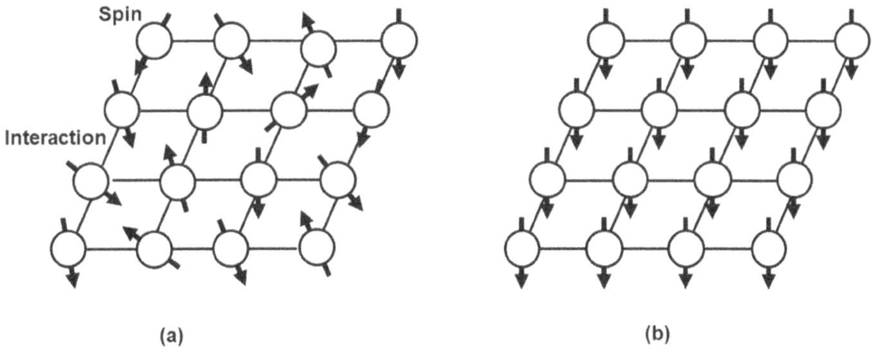

FIGURE 6.3
Ising model of ferromagnetic materials. (a) State at high temperature and (b) state at low temperature.

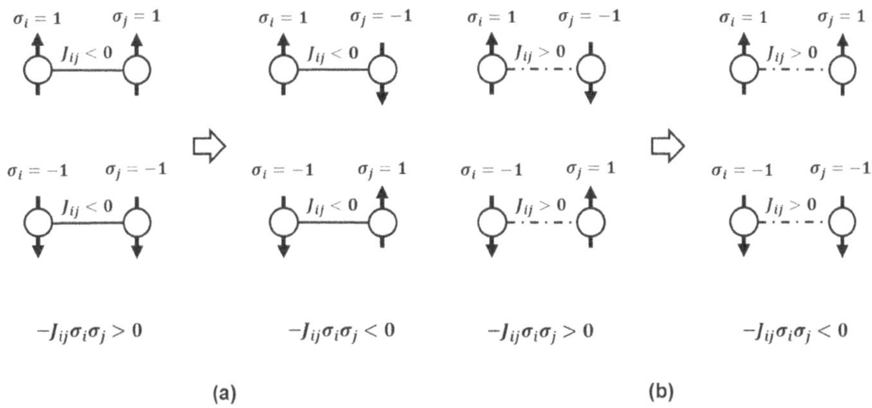

FIGURE 6.4
Interactions and stable spin combinations. (a) Interaction 1 and (b) Interaction 2.

that each take the values 1 and –1, corresponding to two orientations. The same sign represents the same orientation. The interaction (coupling coefficient) between σ_i and σ_j is considered as J_{ij} and is represented by a line in the figure. The spins σ_i and σ_j take values of 1 and –1, respectively, so if we express the energy of the two spins and the interaction between them as $J_{ij}\sigma_i\sigma_j$, then in the case of $J_{ij} < 0$, the energy is lower when the two spins are oriented as shown in Figure 6.4a, and the energy is lower when they face the opposite direction as shown in Figure 6.4b for $J_{ij} > 0$. A mechanism is incorporated that determines the direction of spin based on a balance between this and the tendency for the direction of spin, represented by temperature, to orient in any direction (the higher the temperature, the more likely it is to orient in any direction).

This provides a model for the magnetization of ferromagnets in two dimensions. This can also be achieved by aligning some of the spins upward and some downward in the two-dimensional spread of the lattice points where the spins are arranged. Furthermore, the magnitude of the interaction may be different for every combination of spins. All of these will be discretely oriented at high temperatures, and the states corresponding to the interactions between the spins will emerge at low temperatures.

In this mechanism, even if a disparate spin orientation is initially given, the spin orientation changes to a low-energy state according to the interaction between the spins. From the correspondence with the combinatorial optimization problem, this low-energy state is the solution to the combinatorial optimization problem. This method of applying natural principles to computation is also called natural computing.

6.2.2 Frustration

Since many spins interact with each other, there is in fact competition or imbalance between the spins or between certain regions of spin. This is called a frustrated state. That is, as shown in Figure 6.5a, in the Ising model, spins interact with each other either negatively (in the same direction) or positively (in the opposite direction). Depending on the spin configuration, there is a specific pattern in which the interaction energy is minimized, and at this time, the system reaches a stable state. However, a frustrated state develops when, for example, three or more spins interact, each competing with the other spins. Due to this competition, some of the spins will not be in

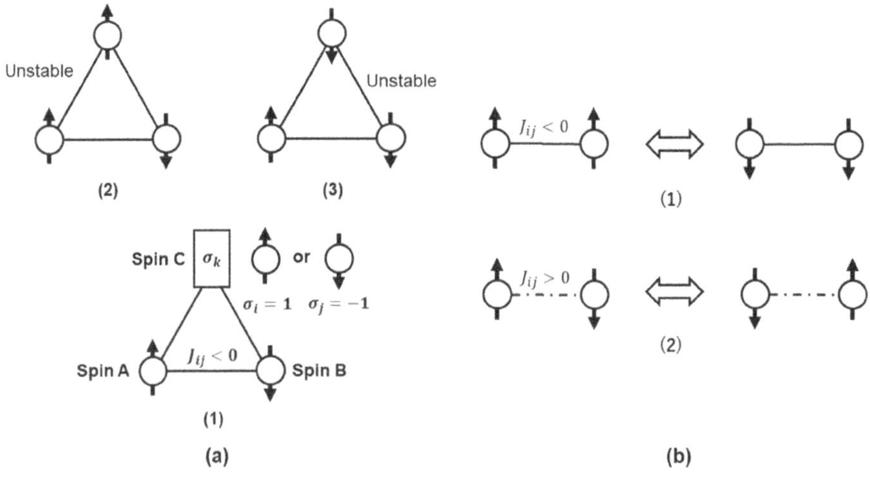

FIGURE 6.5
Frustrated and oscillation states. (a) Frustration and (b) oscillation.

the optimal configuration and will remain in an unstable state. In the example in Figure 6.5a, the three spins (A, B, and C) affect each other, and the interactions between A and B, B and C, and C and A are negative. However, A, B, and C cannot be stable spin orientations with negative interactions at the same time. If A and B have opposite orientations in (1), then A and C are unstable in (2), and B and C are unstable in (3). The frustration state is that none of these spins settle down to a completely stable state. In the actual system, many more spins contribute to the frustration state, and the system as a whole behaves in such a way that it moves to a lower energy state. However, it is necessary to solve the situation in which the convergence of the solution becomes unstable due to the frustration state in the combinatorial optimization problem, which is the subject of this chapter. Even if this problem is solved, when the energy of the solution is observed as the solution progresses, there are problems in which the energy drops uniformly, but there are also cases in which the energy drops discontinuously, as if it were a first-order phase transition in statistical mechanics, affecting the calculation, including frustration. This behavior itself is also a subject of research.

It is important to note that as an actual combinatorial optimization problem calculation mechanism, the Ising machine may include this (a) during frustration, but it is a simpler (b) oscillation. That is, the spins are binary, but two spins can be oriented at the same time and still be energetically the same. This leads to the fact that if there are many spins, the computational process of changing the orientation of two spins at the same time cannot be taken. In particular, if the information processing problem is written in such a way that it can be solved by an Ising machine, the computational process of changing the orientation of two spins at the same time becomes difficult because all the spins are bound together, as described later.

6.3 Combinatorial Optimization Problems and Ising Models

The relationship between the combinatorial optimization problem and the Ising model is shown below [3].

As we have mentioned, there is an element called the spin σ_i that takes two values, and between this pair of spins σ_i and σ_j, we introduce an element of interaction (coupling coefficient) whose coupling strength is J_{ij}. Considering only the interaction between neighboring spins, the model shows the behavior of a two-dimensional magnetic body. Here, we consider a model that includes interactions between arbitrary pairs of spins as a proposal for a combinatorial optimization problem, which we call the fully coupled Ising model. This fully coupled Ising model is also known as the spin glass model [4] in the field of condensed matter physics when long-range interactions are incorporated, although the meaning is different.

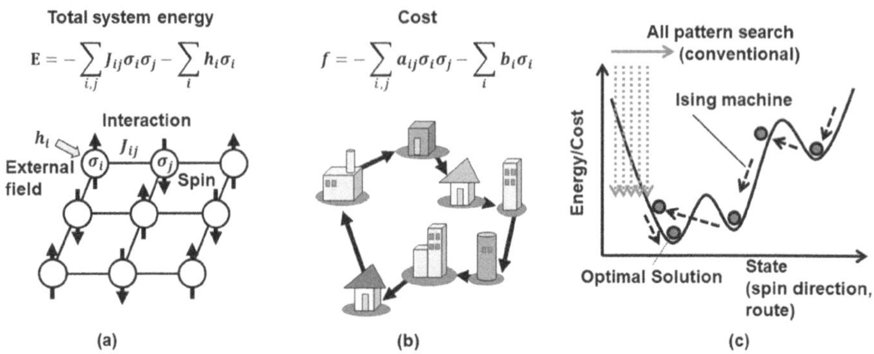

FIGURE 6.6
Ising model and combinatorial optimization problems. (a) Ising model, (b) combinatorial optimization problem (route search example), and (c) solving.

As shown in Figure 6.6a, the expression for the energy of the fully coupled Ising model is where the sums for i, j are taken for all spin combinations.

$$E = -\sum_{i,j} J_{ij}\sigma_i\sigma_j - \sum_i h_i\sigma_i \tag{6.1}$$

Let σ_i be $i = 1, 2, \cdots, n$ and be a component of vector σ, and let h_i be $i = 1, 2, \cdots, n$ and be a component of vector $\boldsymbol{h}$, i.e.

$$\sigma = \begin{pmatrix} \sigma_1 \\ \sigma_2 \\ \vdots \\ \sigma_n \end{pmatrix} \tag{6.2}$$

$$\boldsymbol{h} = \begin{pmatrix} h_1 \\ h_2 \\ \vdots \\ h_n \end{pmatrix} \tag{6.3}$$

Also, the components of the matrix $\boldsymbol{J}$ are J_{ij} $(i = 1, 2, \cdots, n,\ j = 1, 2, \cdots, n)$.

$$\boldsymbol{J} = \begin{pmatrix} \boldsymbol{J}_{11} & \cdots & \boldsymbol{J}_{1n} \\ \vdots & \ddots & \vdots \\ \boldsymbol{J}_{n1} & \cdots & \boldsymbol{J}_{nn} \end{pmatrix} \tag{6.4}$$

Then the energy equation becomes

$$E = -\boldsymbol{\sigma}^T \boldsymbol{J} \boldsymbol{\sigma} - \boldsymbol{h}^T \boldsymbol{\sigma} \tag{6.5}$$

On the other hand, as shown in Figure 6.6b, if we transform the expression for the error function (cost function) f, which is the difference between the solution of the optimal combination and the current combination in a combinatorial optimization problem, the variables here are not defined here, but if we exclude the constant term using a vector or matrix, this energy expression (6.5) has the same form as (6.5). Since this error function f is the difference between the optimal combined solution and the current combination, minimizing it yields the optimal combined solution.

In other words, if we can define the transformation of both variables, we want to minimize the value of the error function in the combinatorial optimization problem, so obtaining the correct result in the combinatorial optimization problem can be read as obtaining the state with the lowest energy in the fully coupled Ising model. Note that equation (6.1) shows that this is mainly the sum-of-products operation that has been performed in neural networks up to the previous chapter.

Then, as shown in Figure 6.6c, if the lowest energy state is obtained by updating the spins so that the energy of the entire system in the Ising model is reduced, the result is the solution to the combinatorial optimization problem. This is the solution to the combinatorial optimization problem. Using the Ising model as an Ising machine to find a solution to a combinatorial optimization problem is much more efficient, i.e., requires less computation, than calculating the errors for all the combinatorial cases, comparing them, and finding the smallest one. In general, the number of combinatorial cases grows exponentially with the number of elements to be combined.

In contrast to calculating all of these, in calculations using the Ising machine as an approximate solution, the number of couplings of fully coupled spins is only the square of the number of spins (corresponding to the number of elements), and these states are calculated while changing the temperature and changing the state of the spins. Counting the total number of calculations of the spin state at a single temperature as one, this converges to a low-energy state in at most tens of thousands of calculations. Compared to the calculation of states that are exponentially proportional to the number of elements, this calculation is much smaller. Furthermore, the Ising machine can be configured in an LSI to reduce power and cost, making it a tool that can be applied at the edge.

The form of the equation being calculated here is the same form as that of the LSI neural network we have seen in the previous chapters, where J and Q are the weights. In other words, it calculates the state when the weight is given, and it becomes a kind of inference process. The fully coupled type is also isomorphic to the Hopfield model. The approach to combinatorial optimization problem seeking can also be discussed in terms of the historical development of neural networks. The study of making Ising machines into LSI chips can utilize the knowledge of neural network LSIs, and conversely, can be expanded to neural network LSIs.

Although the Ising machine is a calculation that yields an approximate solution, it is based entirely on the calculation of the energy in equation (6.1), so the accuracy of the hardware calculation is important for the correct value of the energy itself. The accuracy of the J_{ij} is also related to the exact representation of the problem. In the example of path finding, if we want to find the shortest distance, the accuracy of J_{ij} is directly related to what distance accuracy we want to find. This is the same argument as the accuracy of the weights of neural networks, and in neural networks, there were attempts to make the weights binary. However, this discussion was only possible for neural network weights, including those during learning. When solving optimization problems using the Ising model, J_{ij} is not learned but is determined by the problem itself. This is directly related to the accuracy of the problem setting. Therefore, high accuracy is required. On the other hand, solving optimization problems with the Ising model is an approximate computation, and from this point of view, even if the accuracy is increased blindly, a good practical solution cannot be obtained. 8–10 bits are usually sufficient, although there are special cases where 32 bits or more are required. In some cases, as little as 4 bits are sufficient.

In the Ising model, σ_i is placed as a variable that takes two values, 1 and –1, corresponding to the upward and downward spin. This can be transformed into a single determinant as follows. This formulation of the combinatorial optimization problem is called the quadratic unconstrained binary optimization (QUBO) [5] formulation. Here, we use the variable q_i, which is a transformation of the σ_i variable of the Ising model with the relation $q_i = \frac{\sigma_i + 1}{2}$. It takes the values 0 and 1 as binary values.

The energy equation using this variable is as follows Here, q_i takes only the values 0 and 1, so $q_i = q_i^2$ is used. The components of $i = j$ are summarized here.

$$E' = -\sum_{i \neq j} a_{ij} q_i q_j - \sum_i b_i q_i^2 \tag{6.6}$$

In this form, every term consists of the product of two 0 and 1 valued variables. This is done by first using the following notation:

$$\boldsymbol{q} = \begin{pmatrix} q_1 \\ q_2 \\ \vdots \\ q_n \end{pmatrix} \tag{6.7}$$

$$Q = \begin{pmatrix} b_{11} & a_{12} & \cdots & a_{1(n-1)} & a_{1n} \\ a_{21} & b_{22} & \cdots & a_{2(n-1)} & a_{2n} \\ \vdots & \vdots & \ddots & \vdots & \vdots \\ a_{(n-1)1} & a_{(n-1)2} & \cdots & b_{(n-1)(n-1)} & a_{(n-1)n} \\ a_{n1} & a_{n2} & \cdots & a_{n(n-1)} & b_{nn} \end{pmatrix} \tag{6.8}$$

E' is expressed in the following equation.

$$E' = -\boldsymbol{q}^T Q \boldsymbol{q} \tag{6.9}$$

It can be expressed in this simple form. Finding the vector $\boldsymbol{q}$ for which E' is the smallest is solving a combinatorial optimization problem.

6.4 Hopfield Network and Ising Model

As mentioned in Chapter 4, attempts to solve combinatorial optimization problems by considering the energy of the system as in (6.1) date back to Amari [6] in 1972 and Hopfield's model in 1982 [7] and 1985 [8]. In 1982 paper [7], Hopfield considered a model of a network with symmetric interactions between each of the two values of each neuron shown in Figure 6.7a, where the difference in energy before and after a change in one neuron determines whether it occurs or not.

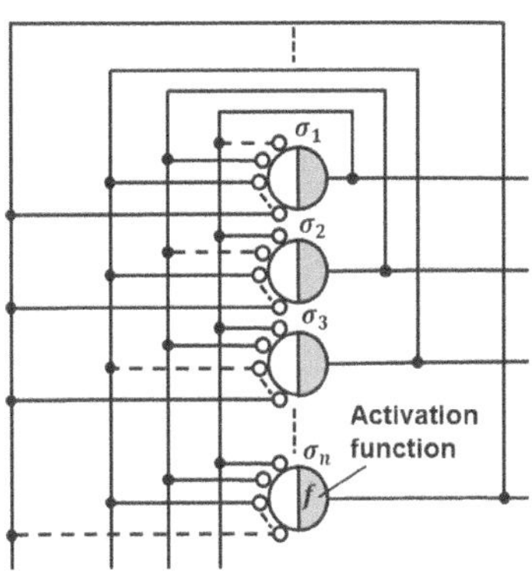

(a)

Activation function f	Network functions
Deterministic	Hopfield network
Stochastic	Fully coupled Ising machine
	Boltzmann machine

(b)

FIGURE 6.7
Hopfield network and Ising machine. (a) Network schematic and (b) activation function network functions.

The relationship between this Hopfield network and the Ising model is described here. We begin our discussion with the energy in (6.1).

$$E = -\sum_{i,j} J_{ij}\sigma_i\sigma_j - \sum_i h_i\sigma_i \qquad \text{(identical to 6.1)}$$

If we focus on one spin σ_i in this equation, the energy contribution to it is

$$\Delta E_i = \sum_j J_{ij}\sigma_j + h_i \qquad (6.10)$$

This is the same as calculating the sum of products of one neuron in the neural network in Figure 6.7a if J_{ij} is considered a weight. The spin σ_i can be regarded as a neuron. Note that equation 6.10 treats the sign of h_i differently from Hopfield's paper, but this can be considered equivalent because of the way h_i itself is valued.

In the Amari–Hopfield model, the value of the neuron or spin σ_i is initially 0, and the σ_i is updated using an activation function that updates it to 1 if the value of ΔE_i is positive (it goes down in energy) and 0 if it is negative. This model was shown to have applications in associative memory by working to estimate some of the spins when they are unknown in the 1982 paper [7], and in a 1985 paper [8], it was shown to give solutions to combinatorial optimization problems.

However, in this method, the value of the neuron or spinσ_i is an activation function that is "determined" to be 1 if the value of ΔE_i is positive and 0 if it is negative. Therefore, when the local energy is small, the state cannot be updated any further. In other words, there was a problem of falling into a locally optimal solution. The Boltzmann machine assumes that the value of the neuron or spin σ_i is probabilistically determined by the value of ΔE_i. At this time, it is possible to incorporate an indicator such as temperature, which is an indicator of the overall fluctuation of the system, and this allows us to escape from the locally optimal solution. The solution itself is the same as the fully coupled Ising machine. The operation of the fully coupled Ising machine itself is described in 6.6.

6.5 Fully Coupled Ising Model

When the combinatorial optimization problem is defined as an error function that represents the difference between the optimal solution and the trial, it is isomorphic to the energy equation that determines the behavior of the fully coupled Ising model. Therefore, minimizing the energy of the fully coupled Ising model was to obtain the optimal solution to the combinatorial optimization problem. For methods to transform various combinatorial

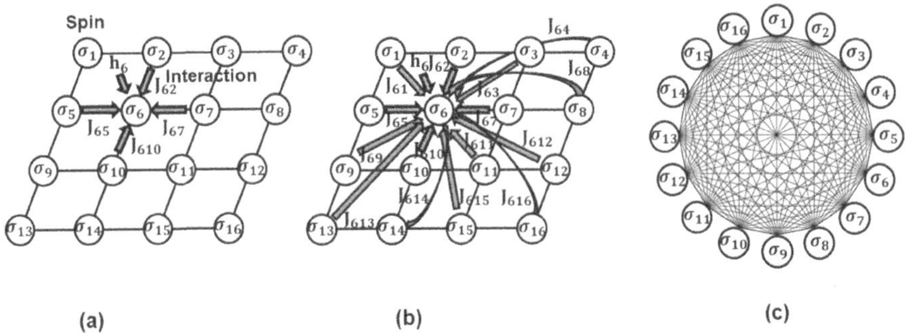

FIGURE 6.8
Sparse coupled type and fully coupled type of spin interaction. (a) Sparse coupled type (σ_6 as an example), (b) fully coupled type (σ_6 as an example), and (c) fully coupled type.

optimization problems into fully coupled Ising, please refer to the literature [3]. Tools have also been developed to automate these methods.

The first model derived from the magnetic model is the adjacent coupled Ising model. It takes the spins at the lattice points as σ_i and interacts only with the neighboring spins. Here, only the interaction at the square is assumed in Figure 6.8a, where the spins are placed at 16 lattice points. If we focus on σ_6 here, there is an interaction between the adjacent spin, $\sigma_2, \sigma_5, \sigma_7, \sigma_{10}$, and $J_{62}, J_{65}, J_{67}, J_{610}$ towards this σ_6, which is indicated by the arrow. It also shows that this σ_6 has an external influence (external field), h_6, by analogy with the model of a magnetic body.

On the other hand, Figure 6.8b shows the case of the fully coupled Ising model. As in (a), we focus on σ_6. This is the model that will incorporate J_{6j}, the interaction from the spins at all other lattice points shown in this figure. The actual error function of the combinatorial optimization problem is isomorphic to this model. Therefore, it is important to use this fully coupled Ising model.

Here, since σ_6 shown in (b) is not special, Figure 6.8c shows the 16 spins and all the couplings between them. This is also Figure 6.7a (they are the same). The fully coupled Ising model with this coupling is the one used to solve the combinatorial optimization problem. It is tightly coupled, i.e., there is a symmetric network of fully coupled interactions between each of the two values of the spins.

The issue for hardware is how to implement this tight coupling in an LSI or whether it is possible to divide it into multiple chips.

6.6 Minimum Energy State Search Method in Ising Model

This section describes a method for finding the minimum value of the energy of the fully coupled Ising model.

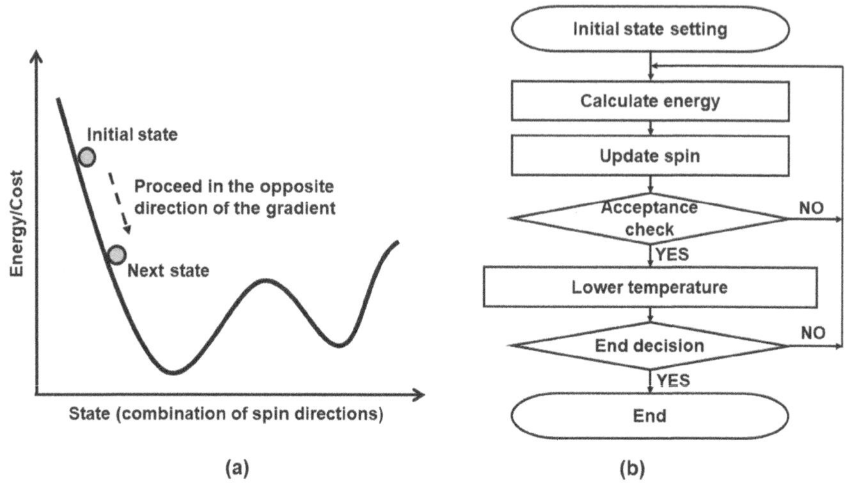

FIGURE 6.9
Searching for the minimum energy state. (a) Diagram of the solution method and (b) simplified flowchart of simulated annealing.

The Ising model with spin number n is a system whose energy distribution is determined from n variables. We want to find the lowest energy state in this system. To do this, we first select an initial state. We then randomly try different states, compare their energies, etc., and if the new state is adopted, we use a method [6] to exchange it with the initial state. This method is called search.

The simplest of these is the mountain-climbing (or mountain-descending) method shown in Figure 6.9a. Here, the current state of energy or cost is represented as a variable. A state is a combination of spin orientations, which can be random at first. From there, the energy is sought for the states in the neighborhood of the state. The change ΔE between these two energies is calculated, accepting the state with the energy in the downward direction, so that it approaches the state with the minimum energy. The gradient is to be updated to a state that is in the opposite direction. This is the same principle as the Hopfield network.

However, this mountain-climbing method can only find local minimum solutions. In the case where there are values with minimal energy for multiple surroundings, i.e., for functions for which local solutions exist, it is not possible to find a solution because it is trapped in a local minimum solution. For this reason, an algorithm is needed to escape from the local minimum solution under some conditions when it is stuck in the local minimum solution. For problems with complex energy distributions, such as spin, it is difficult to converge to a local minimum solution and reach the true minimum energy state.

Therefore, we use a method that adds a stochastic component to this mountain climbing method. This method moves in a probabilistically likely direction, but even if it is trapped in a local minimum solution, it gives it a chance to escape from it probabilistically and find an overall optimal solution. In problems such as the Ising model, a technique called "simulated annealing" [9–11] is used for this purpose. In other words, simulated annealing introduces the concept of temperature in order to find the optimal solution. This method uses a stochastic approach to find the overall optimal solution without sticking to local minimum solutions. This also models the annealing method, in which a clean structure is obtained by gradually cooling a high-temperature metal. Figure 6.9b shows a simplified flow. Here, a high-temperature state is set, and spins are assumed to rotate easily due to this effect, and a high-energy state is also accepted as a stochastic factor. Thus, if a state is locally low energy, it can be transferred to another lower energy state, overcoming the higher energy state in between. This temperature is gradually lowered, making the stochastic change to a higher energy state less acceptable. This prevents a return to the higher energy valley (local minimum solution) and transitions to a deeper valley. Finally, by lowering the temperature sufficiently low, a minimum energy state, or solution, can be reached.

The basic algorithm for simulated annealing is as follows. First, a suitable temperature T *is* given as an initial value. Also, as mentioned above, if we focus on one spin σ_i, the energy contribution to this is as follows.

$$\Delta E_i = \sum_j J_{ij}\sigma_j + h_i \qquad \text{(identical to 6.10)}$$

In other words, when this spin changes, the energy changes by this ΔE_i amount. We define the probability P_i that the spin σ_i changes from 0 in one state to 1 in the next state as follows.

$$P_i = \frac{1}{1+\exp\left(-\frac{\Delta E}{T}\right)} \tag{6.11}$$

By comparing the value of P_i with a separately generated random number to determine whether or not to flip the spin, the spin is flipped with this probability.

If the temperature is high, i.e., $\Delta E \ll T$, then $P_i \approx 0.5$ will be used to randomly determine whether to invert or not. When the temperature is low, the probability of reversal is determined by ΔE. From this, P_i is obtained, and whether or not to invert the spin is determined by comparing this value with a separately generated random number. After performing this operation a certain number of times, the temperature is decreased according to a certain rate. It is known that asymptotic convergence to the optimal solution is achieved by repeating these operations until the temperature decreases completely.

The temperature reduction method is as follows at n in the update step.

$$T_{n+1} = \frac{T_1}{\log n} \tag{6.12}$$

This ensures asymptotic convergence to the optimal solution. However, this takes too much time, so it is more common to use as

$$T_{n+1} = \gamma T_n \tag{6.13}$$

with

$$0.8 \le \gamma < 1 \tag{6.14}$$

Thus, in simulated annealing, the transition probability, represented by the parameter temperature T, determines the transition of the state. As shown in Figure 6.10, while the temperature is high, a wide area is explored by randomly trying out candidate solutions. However, as the temperature decreases, a more localized region is explored in detail to find the optimal solution. In other words, although the distribution is initially uniform, it is gradually changed by varying the temperature while allowing it to converge sufficiently to a steady state. At this time, the influence of probabilistic fluctuations caused by temperature changes reduces the possibility of being trapped in local solutions, enabling global optimization.

Considering integrated circuits, the actual calculations in this simulated annealing method consist of three basic steps: ΔE calculation (inner product calculation), calculation and acceptance determination of spin update values, and temperature calculation (annealing control). Given a combinatorial optimization problem, this is converted into interaction and external field values, which are used to repeatedly calculate ΔE for all spins, calculate new spin values, and make acceptance decisions in a single calculation step, followed by a temperature decrease. The entire calculation step is then repeated until the temperature reaches zero.

There are two challenges in making simulated annealing into integrated circuits.

The first issue is that a random number is required, which must be generated in the circuit, but this random number must have a certain degree

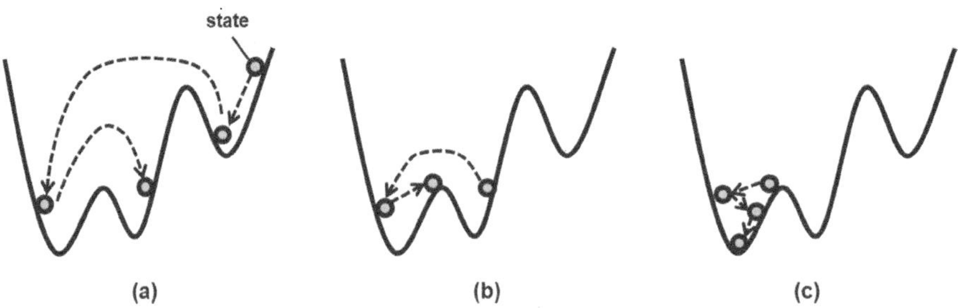

FIGURE 6.10
Searching for a solution using simulated annealing. (a) High temperature, (b) intermediate temperature, and (c) low temperature.

of accuracy. The spin state is determined by comparing the probability calculated by the equation with the random number generated inside the LSI. This is a function similar to the activation function of a neural network, and to take advantage of this nonlinearity, the random number needs to have a precision of 7-8 bits. This needs to be generated using, for example, a linear feedback shift register.

The second issue is the difficulty in realizing the sigmoid function. As explained in the neural network, analog circuits are indispensable to implement the sigmoid function on an LSI without quantization. However, analog circuits have many constraints to be considered in design, and if digital circuits are used in energy calculation, it will be more difficult for analog circuits to use finer CMOS for its high performance. A device to solve these problems is needed. Since there are several ways to solve these problems, an example is given in the next chapter. In the above, the energy calculation was a sum-of-products operation of the spin and interaction values, and the determination of the updated spin is a nonlinear transformation of the result of this energy calculation. In both cases, the neural network innovations described in the previous chapters can also be used.

6.7 Ising Machine

Let us call the hardware configured to solve combinatorial optimization problems using the Ising model an Ising machine. As shown in Figure 6.11, historically, a quantum bit-based Ising machine was announced by D-Wave in Canada in 2011 as Quantum Annealing [12]. This was regarded as the realization of a type of quantum computer, which had been considered a dream technology, and it also had the ability to solve combinatorial optimization problems, which had been a challenge due to the explosive growth of computational complexity. This had a significant impact on the world. Later, it was recognized that quantum technology was not essential for solving combinatorial optimization problems, and in 2015, Hitachi announced a conventional CMOS circuit-based Ising machine called CMOS annealing [13]. This also caused a stir and led to subsequent announcements from multiple institutions, establishing a major trend in this field.

The relationship between "Quantum Gate" and "Quantum Annealing" methods, which are quantum computers, and "CMOS Annealing" methods, which are also called classical computers by contrast, is as follows. A quantum gate system, as discussed briefly in the previous chapter, is a hardware that uses information at the quantum level, called a qubit. Unlike ordinary bits (0 or 1), qubits have characteristics such as superposition and quantum entanglement that are difficult to realize in classical computers.

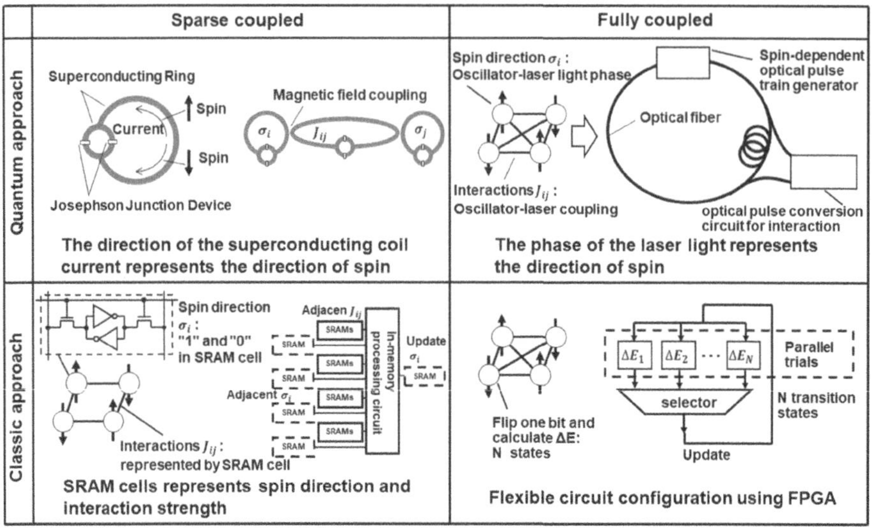

FIGURE 6.11
Historically pioneering examples of Ising machine configurations.

Utilizing these characteristics, quantum gate machines can solve complex problems at high speeds. Among these, they have also been applied to solving combinatorial optimization problems. On the other hand, quantum annealing methods and hardware are specialized for combinatorial optimization problems, and unlike quantum gates, they are not general-purpose. However, for problems using the Ising model, quantum annealing utilizes quantum effects to find the minimum energy state from the energy function of the problem. This makes it possible to find the optimal solution faster for certain problems. The CMOS annealing is a method of solving combinatorial optimization problems using an architecture based on the circuitry of a classical computer. It does not use qubits, but instead performs calculations based on classical bits (0 or 1). It uses ordinary CMOS circuits to search for the optimal solution to the Ising model problem.

While this section provides a brief overview of these published Ising machines, the fully coupled Ising machine with integrated circuits is discussed in detail in the next chapter as a case study of a specific development.

6.7.1 Quantum Ising Machine

Ising machines, which take advantage of the characteristics of quantum systems, were historically developed first.

1. Machine with superconducting coils:
   Spin has two states: upward and downward. This is realized and utilized by the state of the superconducting current flowing

clockwise or counterclockwise in the superconducting ring. The interaction is implemented in the rings placed within them, and the coupling between the spins is called a coupler. It is believed that the tunneling effect can be used to reach the optimal solution at high speed. First, a group of spins is initialized in a quantum superposition state of upward and downward spins by applying a transverse magnetic field. Then, the transverse magnetic field (which is called this because it tilts the spins upward or downward) is gradually weakened to obtain only the spin–spin interactions corresponding to the desired Ising model, thereby utilizing quantum fluctuations to obtain the optimal energy state with high probability. During this process, it is said that the system can quickly reach the optimal solution by escaping from local stable points via quantum tunneling effects. However, superconducting rings are arranged two-dimensionally on a chip, so their interactions are limited to those between neighboring rings, i.e. adjacent coupled machine.

2. Machine with pulsed laser beam:

   Initially, the idea began with the use of a single laser oscillator as an artificial spin, which was then interconnected in a network. However, in the current advanced equivalent approach, multiple light pulse trains, each containing spins and interactions, are first injected into a long optical fiber. Subsequently, portions of the light pulse trains are sequentially extracted, the spin–spin couplings are calculated, and the results are reincorporated into the light pulse trains, enabling the realization of arbitrary mutual couplings. This architecture is known as the Coherent Ising Machine [14] and uses measurements and feedback to achieve full coupling. Due to its unique computational mechanism (utilizing branching phenomena), this system has been reported to achieve higher speeds and accuracy than machines using superconducting coils for problems requiring fully connected networks. Although this system is classified as a quantum system due to the presence of interactions between light pulse sequences, it is sometimes categorized as a classical system. Additionally, this system is considered a pioneer in hybrid systems combining quantum and classical systems.

6.7.2 Ising Machine with CMOS Integrated Circuit

An adjacent coupled Ising machine using CMOS circuitry was announced by Hitachi, Ltd. in 2015. As mentioned earlier, this announcement was received with great excitement. The technology treats SRAM cells as spins, with the SRAM storage information, 0 and 1, corresponding to the two orientations of the spins, and uses electronic circuits to achieve the interaction. In terms of business, although the company states that Ising machines are realized in the cloud, this development has led to the development of Ising machines using CMOS integrated circuits around the world.

As Fujitsu works to improve computing technology to solve social issues, it has developed Ising Machine [15] using CMOS circuits for specialized applications in combinatorial optimization problems and has put the fully coupled type into practical use. First, they developed an FPGA-based system, followed by a dedicated LSI.

Toshiba has introduced a unique calculation method called Simulated Bifurcation Algorithm [16] to obtain the ground state of energy (using branching phenomena), which is reported to achieve two orders of magnitude faster than simulated annealing in fully coupled mode.

6.7.3 Adjacent Coupled Ising Machines and Fully Coupled Ising Machines

1. Adjacent coupled (sparse coupled) Ising machine

   Adjacent coupled Ising machines are historically renowned for both quantum and CMOS systems. The challenge with this technique is that there are cases where the spins *i* and *j* are not hardware-coupled, and the problem must be reformulated to accommodate this. This reconfiguration itself can become a combinatorial optimization problem.

2. Fully coupled Ising machine

   The fully coupled Ising machine is, as its name implies, based on the fully coupled Ising model in which all of the spins *i* and *j* are coupled. The combinatorial optimization problem can be implemented as is. As shown in Figure 6.8, the challenge for this technology is how to implement it on an LSI, since all spins are individually coupled. Although this is an issue for integrated circuits, if this technology can be implemented in an LSI, it will be a better tool for solving combinatorial optimization problems than the sparse coupled Ising machine.

3. Ising machines in the cloud

   This is a method in which a dedicated chip or CPU cluster is used to implement the operations specific to the Ising machine, and this is then used via the cloud. This method can realize a large-scale Ising machine. This method is widely commercialized in Japan, can handle 1 million spins, and be set interactions with 64-bit precision.

It is capable of handling 1 million spins and setting interactions with 64-bit precision.

There is a method to solve combinatorial optimization problems, which require an explosion of computation, in a brute-force manner over all states and at high speed, and this is the quantum-gated quantum computer [17], which is expected to be very promising. This method ultimately uses superposition of states to perform massively parallel computations. If this quantum-gated quantum computer is developed in the future, it will be able to give an exact answer because it calculates directly, rather than indirectly as in the Ising machine.

6.8 Simultaneous Spin Update in Ising Machines

We consider speeding up the operation of the Ising machine. This can be done by calculating the local energy ΔE_i of each σ_i and updating σ_i simultaneously for multiple spins. Theoretically, this increases the calculation speed in proportion to the number of simultaneous updates. However, in the fully coupled Ising machine, when updating one spin, all other spins are referenced. Therefore, simultaneous updates will cause errors in the ΔE_i calculation results, increasing the possibility of updating spins in the wrong direction. This is also an issue for neighboring, but in the fully coupled case, each spin refers to all spins, so it is troublesome.

Consider the case of coupling with four spins at adjacency, as shown in Figure 6.12. The ΔE_6 of the spin σ_6 is as follows:

$$\Delta E_6 = J_{62}\sigma_2 + J_{65}\sigma_5 + J_{67}\sigma_7 + J_{610}\sigma_{10} + h_6 \tag{6.15}$$

Suppose this changes σ_6 to σ_6'.

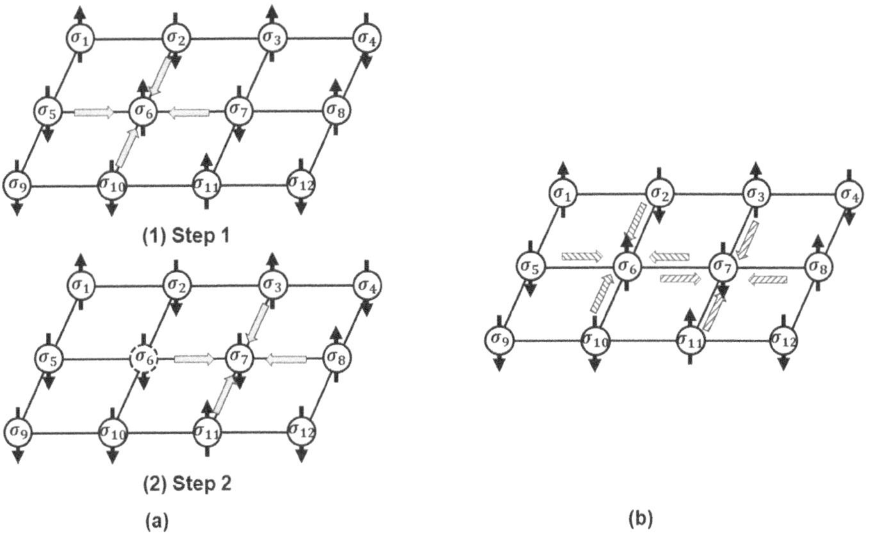

FIGURE 6.12
Challenge of simultaneous updates. (a) Sequential update spin and (b) parallel update spin.

Next, consider the spin σ_7. Here, the issue arises as to whether the relationship with the update of the spin σ_6 is sequential or simultaneous.

In other words, the sequential case is as shown in Figure 6.12a.

$$\Delta E_7 = J_{73}\sigma_3 + J_{76}\sigma_6' + J_{77}\sigma_8 + J_{711}\sigma_{11} + h_7 \tag{6.16}$$

When investigating the dynamics of a system in which spins are mutually coupled, if this is to be examined using simulation, first select one spin at random and determine its influence. Next, select another spin at random and determine its influence in the same way. Continue selecting one spin at random. By doing so, it is possible to simulate the mutual coupling.

However, to improve the calculation speed, if spin σ_6 and spin σ_7 are updated simultaneously as shown in Figure 6.12b, spin σ_6 and spin σ_7 become the following at the same time as the update in Equation 6.15.

$$\Delta E_7 = J_{73}\sigma_3 + J_{76}\sigma_6 + J_{77}\sigma_8 + J_{711}\sigma_{11} + h_7 \tag{6.17}$$

If we look at the spin σ_7, unlike the sequential case, we are using the spin σ_6 instead of the spin σ_6'. Such a relationship occurs with the other spins that we are trying to update at the same time. This does not simulate mutual coupling. Even if the calculation is parallelized, the result will not be correct. Therefore, the only cases in which simultaneous updates are allowed are those in which the spins do not affect each other's orientation, i.e., the interaction is either zero or so small that it does not affect each other. In the case of a tightly bound total coupling, it would be difficult to simultaneously update even two spins.

Therefore, as things stand, the only option is to update each spin one by one (sequential updating), which means that the advantages of integrated circuitry, such as parallel operation, cannot be exploited. A design that is suitable for fully connected dynamics is required.

Of course, since parallel operation is important for high speed, there are many proposals. Table 6.2 lists some of them. Method 1 states that all spins can be updated simultaneously if two sets of all spins are prepared. However, the simultaneous update requires precise control between the two sets. Method 2 uses mean-field theory to approximate the energy difference ΔE_i. This allows all spins to be updated simultaneously. This makes it possible to update all spins simultaneously. The challenge lies in the fact that calculating the approximation of ΔE_i using mean field theory is computationally intensive. Method 3 states that the local energies of the spins for which simultaneous updates are scheduled are precalculated, and the actual updates are performed only for the appropriate spins. This one requires additional circuitry to determine which spins to flip. Method 4 changes the number of simultaneous updates depending on the problem. This requires prior consideration of the distribution of the magnitude of the interaction values, which may be determined by the type of problem, or it may need to be adapted for each actual problem.

TABLE 6.2
Examples of Simultaneous Update Methods

Method			Issue
1	Replica spins	Using replica spins and self-interactions, the spin–spin coupling is transformed into bipartite graph	Self-interaction requires a difficult control algorithm
2	Mean field	Approximates ΔE_i using mean field and continuous-valued spin	Computational cost of ΔE_i is high, and it is difficult to increase the bit width of J_{ij}
3	Prejudgement of update	Updates multiple spins simultaneously within the range, so that the spin update direction does not change	Additional circuits to control the number of spin updates and to determine which spins to flip
4	Based on problem type	Change simultaneous updates degree depend on problems (occasionally to sequential)	Judgment is required for each problem type

6.9 Components of Ising LSI

We will look at a circuit block that constitutes an Ising machine as an integrated circuit with a wired logic control system. As with previous artificial intelligence LSIs, high performance can be obtained by connecting many basic arithmetic circuits.

As shown in Figure 6.13, the basic components of the Ising machine are the part that stores the interaction J_{ij} , the external field h_i, and the spin σ_i, the sum-of-products operator for calculating the spin update, the spin-flip judge, random number generator, temperature calculator, and the controller part for controlling the whole machine. The main contents are as follows.

1. Spin σ_i value registers:

   Since spin takes two values, the circuit that actively represents it and can achieve the smallest area is the SRAM memory cell. An Ising LSI can be built based on this. In addition, a latch circuit can be used to realize the two values of spin, and functions that contribute to the performance of the Ising LSI can be added to the latch circuit, including an SRAM circuit with an initialization function and a function to feedback simulated annealing results.

2. Interaction J_{ij} value registers:

   In the Ising machine LSI, the number of interactions J_{ij} is too large to realize this in a real circuit, so realizing this in a simple circuit is

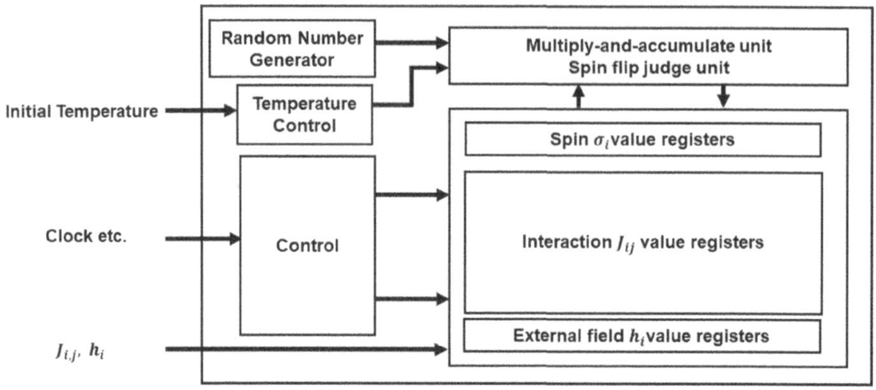

FIGURE 6.13
Ising machine LSI.

important to reduce the overall LSI mounting area. The circuit that can achieve the smallest footprint in an active circuit is the SRAM memory cell. Multiple of these cells are used to achieve the required accuracy (number of bits).

3. Multiply-and-accumulate unit:

   In the Ising machine LSI, the calculation of $\sum_{i,j} J_{ij}\sigma_i\sigma_j$ is the main one. Therefore, only a multiply-and-accumulate circuit is required as an arithmetic circuit. The problem is that the result of the sum-of-products operation has a large number of digits. This becomes an issue when increasing the level of circuit integration.

4. Spin flip determination unit:

   The circuit decides whether or not to flip the target spin based on the set temperature and the result of the sum-of-products circuit. At that time, a random number may be used to update the spin in the direction of increasing energy. This allows the correct solution to be obtained without fitting into a local solution.

5. Random number generator:

   This circuit provides random numbers to the spin flip judging method and is necessary for the Ising machine LSI to obtain the correct solution. When implemented in a digital LSI, it is composed of a shift register and an exclusive-OR (EXOR). This is a pseudo-random number, but it can be made to have a period that is not problematic in practical use. It is also important to generate a true random number from shot noise and thermal noise with as simple a circuit as possible.

6.10 Examples of Combinatorial Optimization Problems in Which Ising Machines Are Applied

As shown in Figure 6.1, the combinatorial optimization problem is extensive. It can be solved by an Ising machine, but must be converted to a form that can be solved by an Ising model.

Since the formulation of the solution by Ising machine has been presented in the previous sections, three examples are given here from the classification as types of problems shown in Table 6.1. Explanations of many other problem examples can be found in the literature [3].

1. Max cut problem:

   A graph can be characterized by a number of vertices (nodes) and edges (edges) that indicate whether the vertices are connected to each other. The strength of the connection between the vertices can be given as a weight to the edge. This would represent, for example, a data structure in which the vertices are individuals and the weights are the strength of the connections between individuals. In this graph, when the vertices on either side of an edge belong to a different set, it is said to be cut. The max-cut problem is the problem of finding the maximal sum of the weights of the edges of a given graph after cutting the vertices that make up this graph into two groups.

   If we denote which of the two groups the vertex i belongs to by the binary value of the spin σ_i and the weight of the edge connecting the two vertices i and j by the strength of the coupling between the spins J_{ij} representing this vertex, this problem becomes a problem of minimizing the following energy E for the two valued vertices and weighted edges that belong in the graph.

$$E = \sum_{i,j} J_{ij}\sigma_i\sigma_j \tag{6.14}$$

   In other words, there is a graph, and J_{ij} is given. In this, we divide the vertices into two, but we can find the maximum sum of weights and the minimum as the above energy in the combination of orientations to the corresponding spins. Thus, it can be solved by an Ising machine. In addition, the number of vertices connected to a single vertex varies from graph to graph. Therefore, for versatility, the Ising machine should be of the fully coupled type.

   In the expression for the energy E, cases with negative and positive signs appear differently for each problem, but this should be determined so that this energy E will be smaller for the solution you seek.

2. Map painting problem:

   This is a problem that requires, for example, coloring a blank map showing the states of the United States so that neighboring states are colored differently. It is known that this can be done in principle with four colors, but in practice, it is somewhat difficult. Using an Ising machine, it is possible to determine the color for each state.

   Since we need two designations for the spin, the state and its color, we denote it with two subscripts: $x_{i\alpha}$. This is one spin, with the first subscript i specifying the state and the next subscript α specifying the color. In terms of U.S. states, there are 51 states, so i has 1 through 51, and if there are four colors, α has 1 through 4. If the same state has a different color, it is represented as a different spin. In this setup, the strength of the coupling between spins J_{ij} is +1 only if neighboring states are the same color. The strength of the coupling between nonadjacent states and adjacent but differently colored states is 0. In the $x_{i\alpha}$, we are going to find, it is 1 if it is realized and 0 if it is not. Therefore, we first need to reduce the following energy E_1.

$$E_1 = \sum_{\alpha \text{ is one of four colors}} \sum_{i,j \text{ are adjacent states}} x_{i\alpha} x_{j\alpha} \tag{6.15}$$

   The strength of the bond between nonadjacent states and states that are adjacent but painted in different colors is zero.

   Also, for a given i (state), only one of the four α (color) has a spin value of 1, and for the other colors, it is 0. Thus, we first have the following.

$$\sum_{\alpha} x_{j\alpha} = 1 \tag{6.16}$$

   Adding this condition means that the next energy E should be smaller.

$$E = \sum_{\alpha \text{ is one of four colors}} \sum_{i,j \text{ are adjacent states}} x_{i\alpha} x_{j\alpha} + \sum_{i \text{ is all states}} \left(1 - \sum_{\alpha} x_{i\alpha}\right)^2 \tag{6.17}$$

3. Traveling salesman problem:

   The traveling salesman problem has come up several times before, but this is a problem of determining the order in which to visit all of the given cities (or locations) so that the total distance between them is minimized. However, there is a restriction that each city must be visited only once and that one cannot visit more than one city at the same time. This is also specified by $x_{i\alpha}$ with the spin having two subscripts corresponding to the cities i and the order α. We want the pair $x_{i\alpha}$ to be 1 if this city is visited in this order and 0 if it is not visited, under the above conditions. Here, let D_{ij} be the distance between two cities i and j. We need to formulate the energy such that the sum of $D_{ij}x_{i\alpha}x_{j\alpha+1}$ is minimized under the following two conditions: one condition is that each city is visited only once, and the other condition is that you cannot visit multiple cities at the same time.

Each city can only be visit once, which is term $\sum_{\alpha}\left(1-\sum_{i}x_{i\alpha}\right)^2$ once, and each city cannot go to more than one city at the same time, as $\sum_{i}\left(1-\sum_{\alpha}x_{i\alpha}\right)^2$. These are added as penalties to the energy formula. Thus, the formula for energy E becomes

$$E=\sum_{i}\sum_{j}D_{ij}\sum_{\alpha}x_{i\alpha}x_{j\alpha+1}+A\sum_{\alpha}\left(1-\sum_{i}x_{i\alpha}\right)^2+B\sum_{i}\left(1-\sum_{\alpha}x_{i\alpha}\right)^2 \quad (6.18)$$

Here, A and B need to be adjusted so that the Ising machine produces a successful solution. The traveling salesman problem is often used to explain this machine because the solution results are illustrated and easy to understand, but it is difficult to obtain an optimal solution due to the adjustment of A and B and the complexity of this formula. Also, although the Ising machine does not obtain an exact solution, the traveling salesman problem tends to make judgments about the obtained solution more stringent because the solution results are easy to understand. For example, when a solution with a few percent difference in path length is judged to be incorrect, increasing the calculation precision of the Ising machine to find a more optimal solution will increase the area of the Ising machine, and complicating the temperature reduction method during annealing will make control more difficult, which may not be optimal.

6.11 Summary

The Ising machine is a specialized computing architecture designed to efficiently solve combinatorial optimization problems that arise in many situations in daily life and business. The number of candidate solutions to these problems increases exponentially with the number of elements, requiring enormous computing resources on conventional von Neumann-type computers. Ising machines can derive solutions to such problems at high speed and low power consumption. Originally, quantum annealing machines were implemented in quantum devices, but later development using conventional CMOS devices became widespread. The computational principle of Ising machines is structurally similar to that of neural networks described in the previous chapter and consists of a sum-of-products operation based on the binary spin state and the interaction between spins (coupling coefficient), and a nonlinear operation to determine spin inversion. In addition, there are two types of spin coupling structures: the "adjacent coupled type," in which only adjacent spins are connected, and the "fully coupled type," in which all spins are coupled to each other. In particular, the fully coupled type can directly express and solve arbitrary combinatorial optimization problems.

References

1. C. H. Papadimitriou and K. Steiglitz, *Combinatorial Optimization: Algorithms and Complexity*, New York, NY, USA: Dover Publications, 1998.
2. K. Huang, *Introduction to Statistical Mechanics*, 2nd ed. New York, NY, USA: Wiley, 1987.
3. A. Lucas, "Ising the Formulation of Many NP Problems," *Frontiers in Physics*, vol. 2, pp. 5–31, 2014. https://doi.org/10.3389/fphy.2014.00005
4. D. Sherrington, and S. Kirkpatrick, "Solvable Model of a Spin Glass," *Physical Review Letters*, vol. 35, pp. 1792–1796, 1975. https://doi.org/10.1103/PhysRevLett.35.1792
5. G. Kochenberger, J. K. Hao, F. Glover, M. Lewis, Z. Lü, H. Wang and Y. Wang, "The Unconstrained Binary Quadratic Programming Problem: a Survey," *Journal of Combinatorial Optimization*, vol. 28, pp. 58–81, 2014. https://doi.org/10.1007/s10878-014-9734-0
6. S. Amari, "Learning Patterns and Pattern Sequences by Self-Organizing Nets of Threshold Elements," *IEEE Transactions on Systems, Man, and Cybernetics*, vol. SMC-2, no. 5, pp. 643–657, 1972. https://doi.org/10.1109/T-C.1972.223477
7. J. J. Hopfield, "Neural Networks and Physical Systems with Emergent Collective Computational Abilities," *Proceedings of the National Academy of Sciences in USA*, vol. 79, pp. 2554–2558, 1982. https://doi.org/10.1073/pnas.79.8.2554
8. J. J. Hopfield and D. W. Tank, ""Neural" Computation of Decisions in Optimization Problems," *Biological Cybernetics*, vol. 52, pp. 141–152, 1985. https://doi.org/10.1007/bf00339943
9. J. Kirkpatrick, C. D. Scott , and M. P. Vecchi, "Optimization by Simulated Annealing," *Science*, 220.4598, pp. 671–680, 1983. https://doi.org/10.1126/science.220.4598.671
10. J. Hajek, "Cooling Schedules for Optimal Annealing," *Mathematics of Operations Research*, vol. 13, no. 2, pp. 311–329, 1988. https://doi.org/10.1287/moor.13.2.311
11. N. Metropolis, A. W. Rosenbluth, M. N. Rosenbluth, A. H. Teller and E. Teller, "Equation of State Calculations by Fast Computing Machines," *The Journal of Chemical Physics*, vol. 21, no. 6, pp. 1087–1092, 1953. https://doi.org/10.1063/1.1699114

12. M. Johnson, M. Amin, S. Gildert, T. Lanting, F. Hamze, N. Dickson, R. Harris, A. J. Berkley, J. Johansson, P. Bunyk, E. M. Chapple, C. Enderud, J. P. Hilton, K. Karimi, E. Ladizinsky, N. Ladizinsky, T. Oh, I. Perminov, C. Rich, M. C. Thom, E. Tolkacheva, C. J. S. Truncik, S. Uchaikin, J. Wang, B. Wilson, G. Rose, "Quantum Annealing with Manufactured Spins," *Nature*, vol. 473, pp. 194–198, 2011. https://doi.org/10.1038/nature10012
13. M. Yamaoka, C. Yoshimura, M. Hayashi, T. Okuyama, H. Aoki and H. Mizuno, "20k-Spin Ising Chip for Combinational Optimization Problem with CMOS Annealing," *ISSCC Digest of Technical Papers*, vol. 24, no. 3, pp. 432–433, 2015. https://doi.org/10.1109/ISSCC.2015.7063111
14. S. Utsunomiya, K. Takata and Y. Yamamoto, "Mapping of Ising Models onto Injection-Locked Laser Systems," *Optics Express*, vol. 19, no. 19, pp. 18091–18108, 2011. https://doi.org/10.1364/OE.19.018091
15. S. Tsukamoto, M. Takatsu, S. Matsubara and H. Tamura, "An Accelerator Architecture for Combinatorial Optimization Problems," *Fujitsu Scientific & Technical Journal*, vol. 53, no. 5, pp. 8–13, 2017.
16. K. Tatsumura, A. R. Dixon and H. Goto, "FPGA-Based Simulated Bifurcation Machine," *29th International Conference on Field Programmable Logic and Applications (FPL)*, pp. 59–66, 2019. DOI: 10.1109/FPL.2019.00019
17. M. A. Nielsen and I. L.Chuang, *Quantum Computation and Quantum Information*, 10th Anniversary ed. Cambridge: Cambridge University Press, 2011.

7

Fully Coupled Ising Machines: A Case Study

In this chapter, we will look at an example of an fully coupled Ising machine LSI configuration.

As shown in the previous chapter, the calculations required by the neural network LSI and those required by the fully coupled Ising machine LSI have many things in common. This is because the processing is similar to that of the sum-of-products operation, and active function is also the main subject. The fully coupled Ising machine LSI can be regarded as an LSI that incorporates an annealing mechanism specialized for combinatorial optimization problems in the Hopfield network, which is a neural network.

7.1 Fully Coupled and Adjacent Coupled Types

As described in Chapter 6, there are two types of couplings between spins in the Ising machine: the adjacent coupled type, which considers only couplings between neighboring spins, which is known as the model for magnetic bodies, and the fully coupled type [1,2], which considers all the couplings of all spins apart from the magnetic model. Table 7.1 summarizes this comparison [3].

Adjacency coupling means that there is only coupling between adjacent spins. Thus, for example, in a typical combinatorial optimization problem called the traveling salesman problem, a basic circuit element representing 1 million spins is required to solve a problem involving 33 cities. In addition, elements that realize interactions with 4–8 neighboring spins that are in close proximity to each spin are also required. In this example, 4–8 million elements are needed. In addition, the mathematical form of the combinatorial optimization problem must be converted so that it can be handled in this adjacency-coupled form.

On the other hand, the fully coupled type can handle the mathematical form of the combinatorial optimization problem as is. Also, only 1,000 spins are needed to solve the same 33-city problem. Since it is fully coupled, the spins need to be connected by 1 million elements to interact with each other. Therefore, to handle the same problem scale, in terms of the number of elements, the adjacent coupled type requires 4 to 8 times more elements than the fully coupled type.

DOI: 10.1201/9781003599067-7

TABLE 7.1

Adjacent/Sparse Coupled and Fully Coupled

Method	(a) Adjacent/Sparse Coupled	(b) Fully Coupled
Schematics	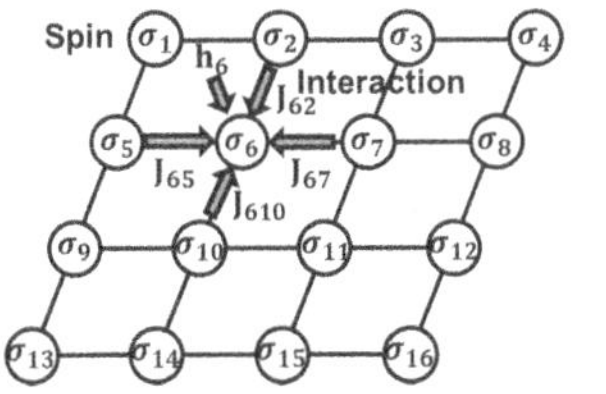	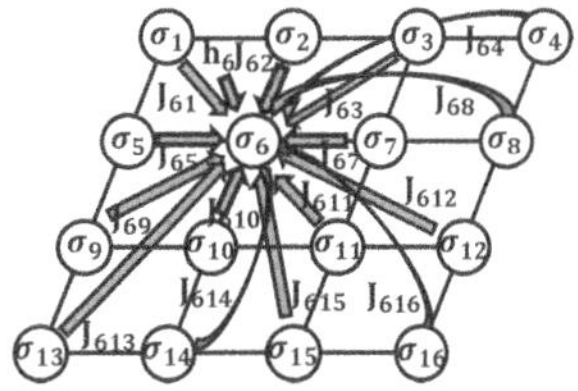
Available problem scale	Small (Example: For 33 cities, 1M spins required. Inferior for solving large-scale problem)	Large (For 33 cities, 1k spins are sufficient. Good for solving large-scale problems)
Versatility, usability	Narrow (annealing format conversion+individual special mapping to hardware for each problem)	Wide (annealing format conversion only)
On-chip spin connectivity	Good (because of adjacent coupling, there are few spin-to-spin bonds)	Difficult (because of full coupling) → Good (with this chapter solution)
Multichip expandability	Good (because of adjacent coupling, there are few joints between chips)	Difficult (because of full coupling) → Good (with this chapter solution)

But on the other hand, wiring in the fully coupled form appears to require that one spin be connected to 999 other spins via interactions. Unlike the software description, how can this be solved physically?

Based on the above, there have been many developments in both areas, but in this chapter, we will focus on development examples that emphasize the high versatility of fully coupled types, namely, their ability to handle the mathematical form of combinatorial optimization problems as they are. As mentioned earlier, the challenges of fully coupled types are, as shown in Table 7.1, that due to their fully connected nature, the number of connections is one-to-many, both on and between chips, and that this number becomes enormous. This also leads to wiring implementation issues.

Based on the above, there have been many developments in both areas; however, this chapter focuses on development examples that emphasize the high versatility of fully coupled types, namely, their ability to handle the mathematical form of combinatorial optimization problems as they are. As mentioned earlier, one of the challenges of fully connected architectures, as shown in Table 7.1, is that due to their fully connected nature, the number of connections – both within and between chips – is one-to-many and becomes extremely large. This also poses issues related to wiring implementation.

7.2 Basic Configuration for LSI of Fully Coupled Ising Machine

First, we describe a scheme that couples all the spins together in a single chip. This allows us to deal with the mathematical form of the combinatorial optimization problem as it is.

7.2.1 Basic Configuration

Once again, the energy of the entire system in the Ising model is expressed by the following equation as described above.

$$E = -\sum_{i,j} J_{i,j}\sigma_i\sigma_j - \sum_i h_i\sigma_i \tag{7.1}$$

The Ising model is used to seek solutions by updating the spins σ_i and σ_j so that E becomes smaller. The method of updating the spin so that E becomes smaller is the simulated annealing method described earlier. In this section, we will first discuss the method devised to compute the equation (7.1). The challenge was that, as shown in Table 7.1b, a large number of couplings were needed because they are all couplings.

First, let us consider a method in which all interactions (coupling coefficients) are represented only by the interaction between the most adjacent spins. Then, as shown in Figure 7.1a, there is a way to prepare an upper and lower spin layer that takes all interactions into account. That is, for each spin, prepare the same number of copies as the number of spins. This creates a two-dimensional array-like arrangement. This is the upper spin layer, and in the array of lower spins, the upper layer is rotated by 90 degrees. By having an interaction between the upper and lower layers here, it is possible to represent everything with only the interaction between the most neighboring layers. Each upper spin σ_i receives interactions from the lower spin σ_j, and all the interactions $J_{i,j}$ received by the same upper spin connected to it are added together to determine the final state and update the lower spin σ_j. On the other hand, if a circuit is implemented using this method on an actual circuit, the wiring distance of the circuit increases as the number of spins increases, including in the case of a three-dimensional structure, and the number of wires also increases. This leads to an increase in parasitic capacitance and an increase in wiring cost.

If we look at the row-by-row operation in Figure 7.1a, we see that σ_1 is updated by $\sum_j J_{1j}\sigma_1\sigma_j$, σ_2 is updated by $\sum_j J_{2j}\sigma_2\sigma_j$, σ_3 is updated by $\sum_j J_{3j}\sigma_3\sigma_j$, ... and so on, one row at a time. In other words, when one row at a time, we can see that we only need to arrange the spins and connect them, the spin interactions, and the arithmetic circuit in a vertical line. Also, for the calculation of the next line, we only need to switch the interaction part.

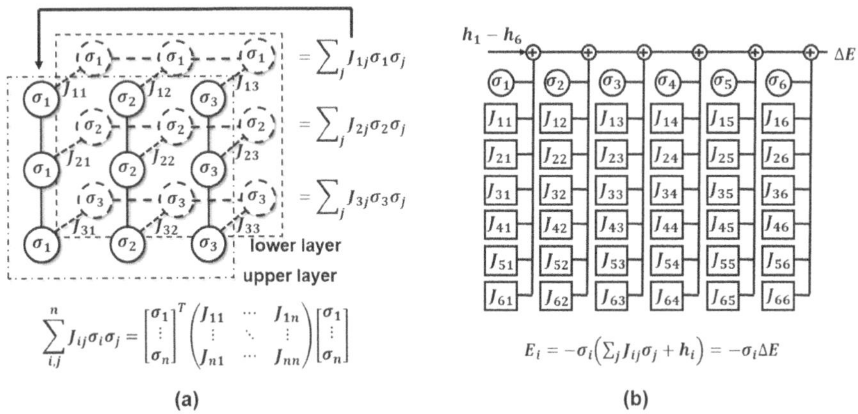

FIGURE 7.1
Fully coupled configuration of Ising machine LSI. (a) Full coupling in two-layer spins and (b) One-dimensional spins+two-dimensional interacting array arrangement.

Reconstructing the scheme based on this, we can see that the spins should be arranged in one dimension and the spin interactions should be arranged on a two-dimensional array, as shown in Figure 7.1b, and each spin interaction (coupling) and calculation circuit should be connected in a vertical line. This allows a large reduction in the size of the circuit and a large number of spins to be loaded in a fully coupled type calculation.

This is the same as the basic structure of AI chips as described so far. It is important to accelerate computation by arranging computing elements in a two-dimensional array, and in-memory computers have developed from a similar concept.

7.2.2 Simulated Pseudo Annealing Method

Next, the spin update method is described.

First, if we focus on a spin σ_i, the energy E_i for this spin is as follows:

$$E_i = -\sigma_i \left(\sum_j J_{i,j}\sigma_j + h_i \right) \tag{7.2}$$

whereas,

$$\Delta E_i = \sum_j J_{i,j}\sigma_j + h_i \tag{7.3}$$

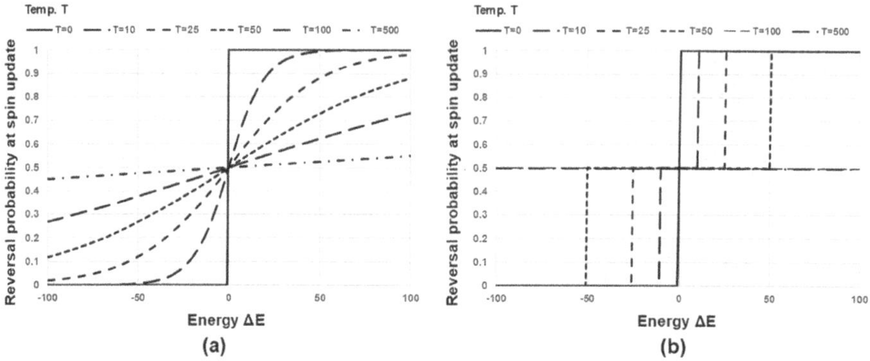

FIGURE 7.2
Comparison of update probability behavior in simulated annealing methods. (a) Conventional and (b) Pseudo annealing.

If we put, this would indicate the change in energy caused by updating the spin. Using this ΔE_i, we introduce the temperature and update the spin based on the following sigmoidal function

$$\sigma_i = -\mathrm{sgn}\left(\frac{1}{1+e^{-\frac{\Delta E_i}{T}}} - r\right) \tag{7.4}$$

where r is a random number between 0 and 1. This allows for stochastic behavior. This is called the simulated annealing method [4] mentioned in the previous chapter, and the probability of spin reversal with respect to energy is shown in Figure 7.2a.

However, sigmoid functions and high-precision random numbers are required. These, including the use of analog circuits, result in a larger circuit mounting area and more stringent operating conditions for the LSI.

Therefore, the example in this chapter is simplified as follows. This is called the Simulated Pseudo Annealing (PA) method [3]. Returning to equation (8.2), this can be viewed as

$$\frac{\partial E_i}{\partial \sigma_i} = -\left(\sum_j J_{i,j}\sigma_j + h_i\right) = -\Delta E_i \tag{7.5}$$

σ_i takes only two states of ± 1 , so updating σ_i so that σ_i and the parentheses, i.e. ΔE_i, have the same sign, will result in a decrease in energy. In other words, we are back to the Hopfield model in a sense. However, if this is not done, we may fall into a local solution. Therefore, we introduce a parameter

T corresponding to the temperature as follows where ± is a random number with only two values.

$$E_i = -\sigma_i \left(\sum_j J_{i,j}\sigma_j + h_i \pm T \right) = -\sigma_i (\Delta E_i \pm T) \tag{7.6}$$

Then, only the sign of this $\Delta E_i \pm T$ determines the update of σ_i.

$$\sigma_i^{\text{NEW}} = -\text{sgn}(\Delta E_i \pm T) \tag{7.7}$$

This is called the pseudo-annealing scheme. Binary random numbers can be generated as pseudo-random numbers using a linear feedback shift register as described in Chapter 3. This eliminates the need for an analog circuit.

The algorithm for this pseudo annealing method is as follows.

Initializing

Generate initial the spin-stateσ and temperature T

// --------------------Pseudo Annealing----------------------

While $MC\ steps < Loop\ Num$

For $i == 0$ to N do //(N is number of spins)

$\Delta E_i \leftarrow \sum_j J_{ij}\sigma_j + h_i$

p = *random number* //(0~1)

if $p<0.5$ then

$T_{calc} \leftarrow T$

else then

$T_{calc} \leftarrow -T$

end if

$\sigma_i' \leftarrow -\text{sign}(\Delta E_i + T_{calc})$

end

$T \leftarrow T * temp_{coefficient}$ $//temp_{coefficient} < 0$

$MC\ steps \leftarrow MC\ steps + 1$

end While

The method of solving combinatorial optimization problems initiated by Hopfield, by considering the energy of the network of all couplings, has evolved into a form suitable for LSI applications.

7.3 Example of Fully Coupled Ising Machine LSI Chip Configuration

The overall structure of the LSI is shown in Figure 7.3. There is an interaction part (2-dimensional array) and a spin part (1-dimensional array), which will be described in the next section, a part that calculates equation (7.6) using pseudo annealing, and a circuit that determines whether or not to update spins according to equation (7.7), and a circuit that controls these parts.

7.3.1 Separated Array Structure in Fully Coupled Type

The main part of the calculation shown in Figure 7.3 is also shown in Figure 7.1b. In this separate array structure in a fully coupled type, the coupling cells are arranged in a two-dimensional array, and the spin cells are arranged one-dimensionally, separated from the coupling cells [3].

In this configuration, data are read from the coupled-cell 2D array in rows. A parallel sum-of-products operation is performed using this and the spin cell data, and a stochastic calculation is performed using temperature and random numbers. A pseudo-annealing method is used to update the spin state using only the sign of the result of this calculation. By repeating this process, the state of multiple spin cells, which is the solution, can be easily determined, resulting in a fully coupled type. Because of the fully coupled

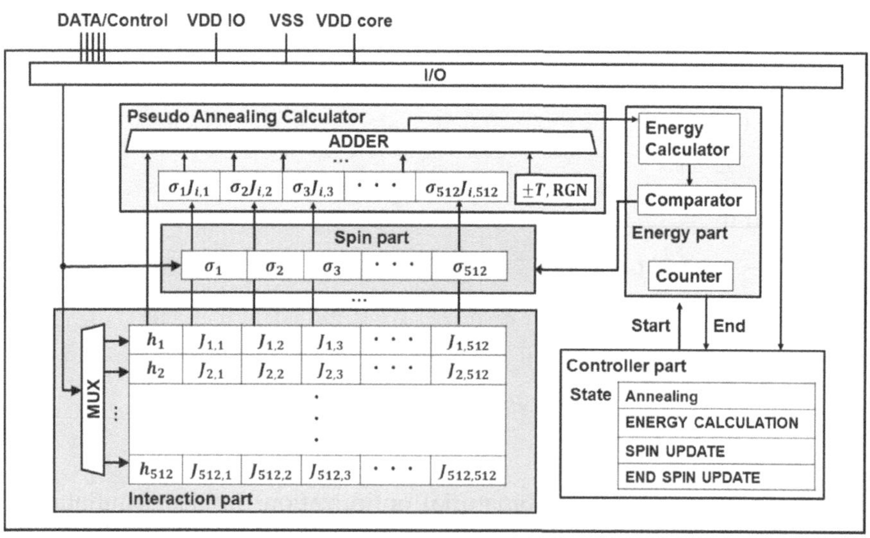

FIGURE 7.3
LSI configuration.

type, the number of spin cells is reduced to the square root of the number of adjacent spin cells, and the number of coupled cells is also reduced to a fraction of the number of adjacent spin cells.

7.3.2 Interaction (Coupling Coefficient) Cell Placement Scheme

When using this method to create LSIs, a little ingenuity is required for the coupled-cell 2D array.

When implementing the J_{ij} in a circuit, from the standpoint of wiring regularity and ease of use, it is sufficient to use the two-dimensional array shown again in Figure 7.4a. However, the energy of the interacting part does not change when the spins are swapped. In other words, since J_{ij} is symmetrical, only half of the J_{ij} parts are needed. This is obvious from an architectural point of view. However, a problem arises when we consider that the entire system should not be constructed by storing the values of J_{ij} in the memory section and connecting it to the arithmetic processing section with a bus, but by using wired logic, which is superior in low power consumption and high speed. In other words, if half of these J_{ij} are removed as they are, the layout would simply be triangular in the two-dimensional array concept.

Therefore, as shown in Figure 7.4b, a portion of this triangle is cut out and moved so that it is folded, making the layout of the entire joined cell array rectangular [3]. Although the composition of the J_{ij} itself is somewhat

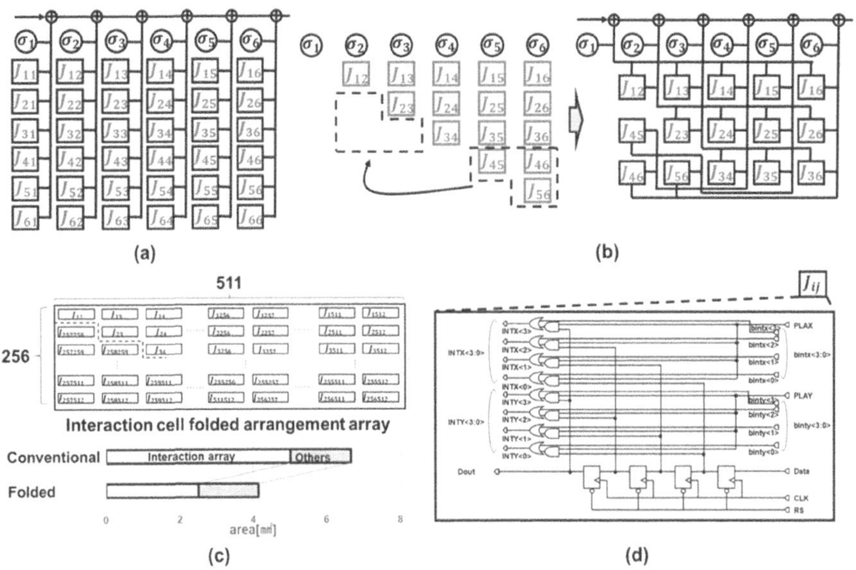

FIGURE 7.4
Interaction cell placement method. (a) Separate array configuration, (b) interaction cell folded arrangement, (c) area comparison, and (d) example of interaction cell circuit configuration.

complicated because of the need to take out more than one, this method is taken because the number is half of the total number.

In an actual LSI example with 512 spins, halving the J_{ij} from 512 × 512 to 512 × 256, the mounting area including the spins was reduced to 62%, as shown in Figure7.4c. An example configuration of J_{ij} with multiple take-outs is shown in Figure 7.4d.

Note that it may not always be necessary to halve J_{ij} when applying in-memory computing techniques. In-memory computing has excellent area efficiency, making J_{ij} with duplicates easier to control.

7.3.3 Spin Thread Method

The separation of the 1D spin cell portion from the coupled cell portion of the 2D array creates a new advantage.

Looking again at the operation of the Ising machine, first, the problem of combinatorial optimization can be represented by a pair of J_{ij}. This is placed in a two-dimensional coupled cell. This was then used to determine the orientation of each spin in the one-dimensional spin cell from equation (7.7). Then, we notice that if we set up multiple spin cell sections (multiple rows) in this 1D spin cell section, the calculations between these multiple pairs of 1D spin cells and the 2D coupled cells can be performed independently in each pair. In other words, it is as if multiple times or multiple chips' worth of operations can be performed at once. This method is called the spin thread method [3], and Figure 7.5 shows an example of its configuration.

This allows multiple calculations to be performed at once, even though the data are transferred in and out of the spin cell portion and the coupled

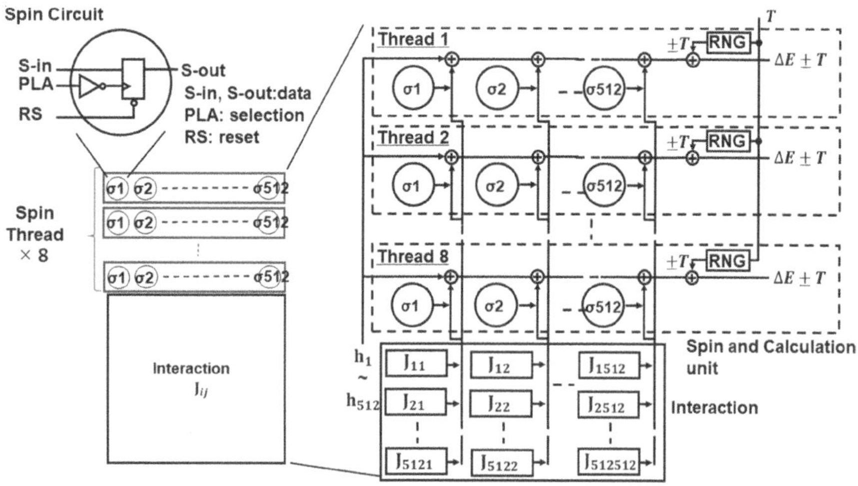

FIGURE 7.5
Spin thread method.

cell portion of the two-dimensional array from the outside world only once. In this figure, eight threads are prepared to update the spins by calculating the energy from the state of the spins at that time using the coupled cells of the two-dimensional array. Thus, this chip can perform eight calculations at once. There is no limit to the number of these threads.

Here, the solution is not uniquely determined because the annealing method is a stochastic calculation using temperature and random numbers. Also, the solution differs when the temperature is changed. The spin-thread method makes it easy to perform calculations that take advantage of these characteristics. For example, independent temperature settings are possible for each spin thread, and the reported chip has this feature. The reported chip has this feature, which requires fewer iterations and allows for different temperature settings for each thread. Furthermore, after a certain number of advances, the spin state of the thread with the lowest energy at that stage can be copied to the remaining threads and advanced. This can increase the acceleration effect of parallel computation.

7.3.4 Fully Coupled Ising Machine

Figure 7.6 shows a photograph of a fully coupled Ising machine LSI [5] created using the technology described so far. It has 512 spins and 8 spin threads. 22 nm CMOS technology was used, and the chip size is $3 \times 2\,mm^2$. The power supply voltage is 1.8 V for the IO system and 0.8 V for the LSI circuit core. The circuit core part operates at 10 MHz and consumes 44 mA. Figure 7.7 shows the measurement system, and Figure 7.8 shows the results of solving the max-cut problem using this system. The problem was generated with 512 vertices, and Figure 7.8 shows the actual measured energy transitions for each thread of the eight spin threads. The computation step on the horizontal axis is defined as one step to update all 512 spins. The minimum energy is

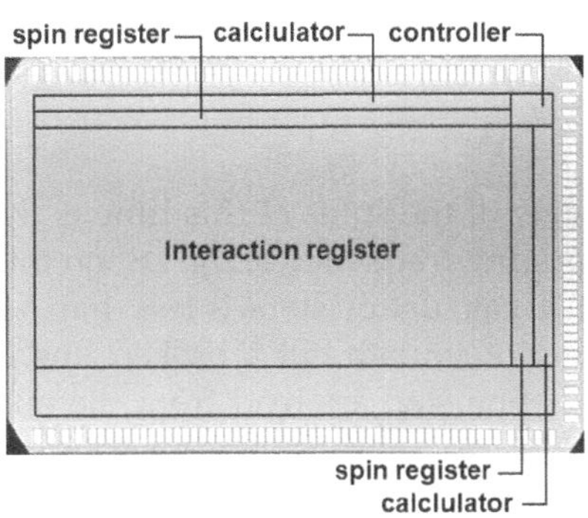

(a)

Technology	22nm CMOS
Chip Size	3mm × 2mm
Op. Freq [MHz]	10MHz
VDD [V] (I/O / Core)	1.8 / 0.8
Op. Current [mA] (I/O / Core)	44 / 1
Op. Power [mW] (I/O / Core)	35.2 / 1.8

(b)

FIGURE 7.6
Example of a prototype chip for a 512-spin fully-coupled Ising machine. (a) Chip photo and and (b) characteristics.

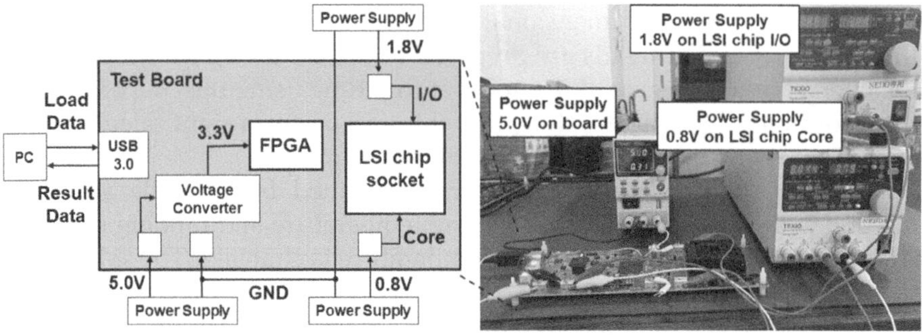

FIGURE 7.7
512-spin fully coupled Ising machine LSI measurement system.

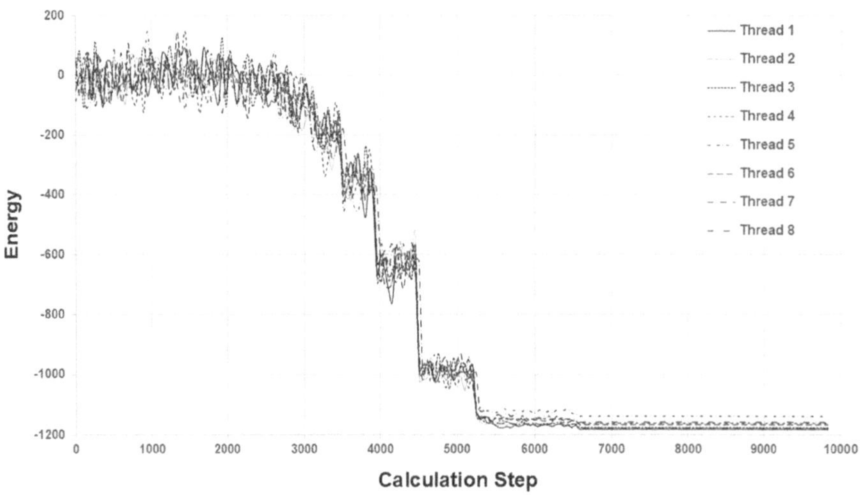

FIGURE 7.8
512-spin fully-coupled Ising machine LSI operation results (actual energy transition measurements).

reached after about 7,000 times, and the state of the spins at this time is the solution to this problem. Although the operating frequency of the LSI on this board is 10 MHz, the time required for 10,000 calculation steps is less than 5 s.

The reason why the energy transition is not smooth but a broken line is that not all states of each step of the calculation step are read from the outside. Nevertheless, as shown in this figure, the energy transitions have jumps. Sometimes, we can see the spin renewal continue for a certain number of times, followed by a transition to a lower energy state. Since the temperature is lowered at each calculation step, the specific heat in this definition is also jumpy, so to speak. Also, each thread has a different value, although they are almost the same. This behavior of the Ising machine is also a subject of research.

7.4 Scalability Configuration Example

There is a limit to the number of spins that can be implemented in a single LSI due to the cost of chip area, which limits the scale of problems that can be solved. The number of spins must be increased to solve large-scale problems, but this is difficult to do with the fully coupled type when the number of spins is increased by using multiple chips, even when the chips are divided into two, as shown in Figure 7.9a. The wiring between the two chips needs to be as many as the number of intersections between the line connecting the spins in this figure and the thick line separating the two chips. The problem is that the number of interconnections between two chips is as many as the number of intersections between the line connecting the spins in this diagram and the two separated thick lines.

In this section, we present a scalable method for constructing a large-scale Ising LSI system [6] that combines multiple fully coupled Ising LSI chips with a very small number of chip-to-chip connections and operates as a single fully coupled system using an annealing method, despite its simple structure as one solution method.

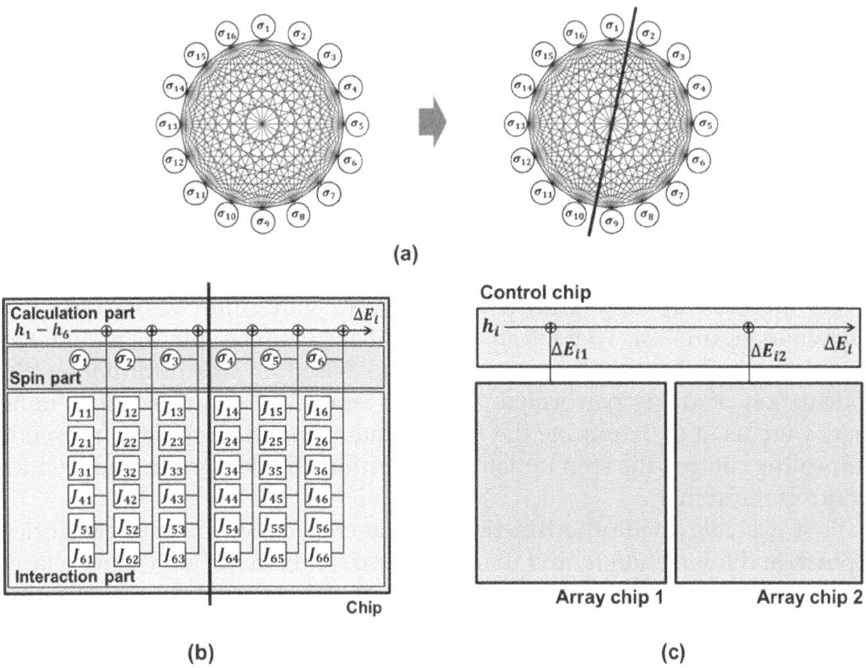

FIGURE 7.9
Scalable method concept. (a) Issues with two-chip configuration, (b) single-chip method, and (c) scalable method.

7.4.1 Scalability

What is shown here is a method of updating the spins by splitting the energy calculation. The concept is to use two different chips, the same number of first chips to split and calculate the energy, and a second chip to sum them up and update them. The whole system works as a single fully coupled system. This method requires very little data communication between the first and second chips. Although it is a simple method of dividing the energy calculation, it enables a scalable fully coupled system as an annealing method.

Figure 7.9b, again as a redescription, is a scheme in which the spins are arranged in one dimension and the spin interactions are arranged as a two-dimensional array. Now $\Delta E_i = \sum_j J_{ij}\sigma_j + h_i$ values were used to update σ_i. Figure 7.9b shows that, unlike Figure 7.9a, this one can be easily divided into two. Figure 7.9c shows the case where the chip is divided into parts. If the spin σ_i is to be updated with ΔE_i, we note that only this ΔE_i needs to be exchanged in multiple chips, even if the number of bits in the interaction $J_{i,j}$, which determines the accuracy, is increased. In other words, to update σ_i, ΔE_i is divided between Array chip 1 and Array chip 2, and ΔE_{i1} is computed in Array chip 1 and ΔE_{i2} in Array chip 2. These are simply scalar quantities. These should be summed in Control chip to obtain ΔE_i. With this ΔE_i, we can decide whether or not to update σ_i and return the binary information for this to Array chip 1 and Array chip 2. This method makes it easy to expand to multiple chips while keeping all the couplings. In principle, there is no limit to the number of chips that can be extended, although it depends on the number of spins.

7.4.2 Example of Scalable Configuration

The following is a step-by-step explanation of specific configuration examples.

The architecture of the fully coupled annealing machine was, to repeat, as shown in Figure 7.9a. There is an element that stores the spin σ_i, the interaction (coupling) J_{ij}, and the external field h_i. Based on this information, the calculation of ΔE is performed. The ΔE, temperature, and random numbers were used to determine the spin update value and decision. In pseudo annealing control, the spin update is determined by the sgn function, which controls the spin σ_i.

First, we can divide the functions of the chip into two main categories. Spin σ_i and interaction J_{ij}, and the rest. Spin σ_i and interaction J_{ij} have a large number of elements, and it is difficult to fit these parts into a single chip when aiming for a large scale.

As shown in Figure 7.10a, these are divided into two chips, the Array chip (A-chip) and the Control chip (C-chip), where ΔE is calculated in the A-chip, and after this calculation, the calculation results from the A-chip are sent to

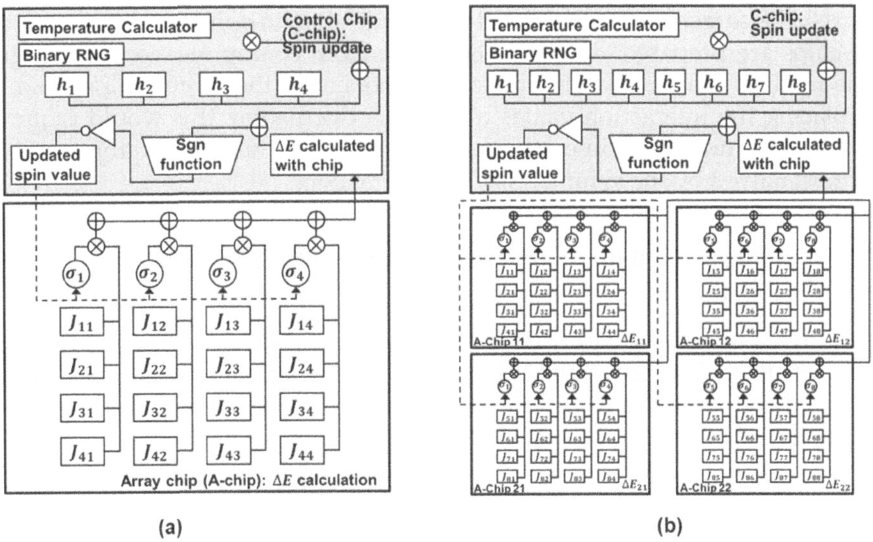

FIGURE 7.10
Scalable configuration of fully coupled Ising machines. (a) Fully coupled Ising machine with functions divided into chips and (b) scalable fully coupled Ising machine.

the C-chip. The C-chip calculates the spin update value and makes an acceptance decision. The C-chip also controls the entire annealing process. In this configuration, the calculation process remains the same even when the number of A-chips is increased. Therefore, by increasing the number of A-chips, the number of C-chips remains one, and the circuit scale can be increased. This configuration is shown in Figure 7.10b.

In Figure 7.10b, the calculation of ΔE is divided into multiple A-chips, and the calculation consists of an additional process of finding the sum of the two in the C-chip. This calculation is shown in the equation (7.8). The calculation of the spin update value at this time is shown in equation (7.9).

$$\Delta E_i = \Delta E_{A\text{-chip}(i,1)} + \Delta E_{A\text{-chip}(i,2)} + \cdots = \sum_{j}^{n} \Delta E_{A\text{-chip}(i,j)} \tag{7.8}$$

$$\sigma_i = -\operatorname{sgn}\left(\Delta E_i \pm T\right) \tag{7.9}$$

In the equation (7.8), the operation is similar to the operation in the conventional architecture, where the interactions in a horizontal row are read, and the calculation of ΔE between chips is also divided into one calculation with chips in a horizontal row, and the value of ΔE required for the spin value update is obtained by finding the sum of the two. Furthermore, in this architecture, in order to multiply the number of spins by n (or divide them into n

parts), the same spin is duplicated to include all spin-spin couplings, and n^2 A-chips are prepared. This is to ensure smooth communication. Although the number of chips can be made proportional to the number of spins by replacing the interaction values during the calculation, this would require extra communication time for the replacement. The number of chips can be almost halved by the symmetry of the interaction.

7.4.3 Example Implementation of a Scalable Fully Coupled Ising Machine

Figure 7.11 shows a board that is an example implementation of a scalable, fully coupled Ising machine. In Figure 7.11, the A-chip described above is divided and implemented in 16 FPGAs, and the C-chip is implemented in one FPGA [6]. The former is called A-FPGA and the latter C-FPGA spin. The board also contains an FPGA called B-FPGA for communication with a PC, etc.

As mentioned above, n times the number of A-chips was n^2 prepared, so the number of spins is four times the number of spins implemented in one FPGA. These work as a single fully coupled Ising machine. The FPGA used in Figure 7.11 has 96 spins per FPGA, making it a fully coupled Ising machine with 384 spins. It can also operate as a 384-spin×2 fully coupled Ising machine with this board configuration by taking into account the symmetry of the interaction.

Figure 7.12 shows the calculation procedure and flowchart. Figure 7.12a is step 1 (problem loading part), where the problem data mapped to the Ising model is sent from the PC to each of the 16 A-FPGAs and 1 C-FPG on the board. The connection to the PC and clock management are performed by B-FPGA, which controls the entire board.

The calculation part of step 2 is shown in (b). When the calculation starts, the A-FPGA calculates the interaction and spin values, respectively, based on

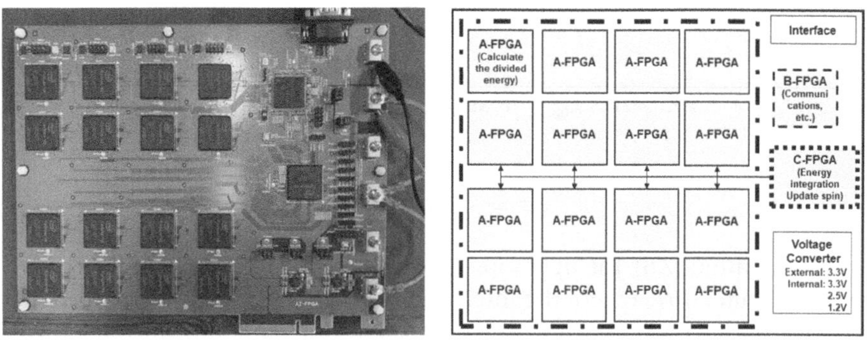

(a) (b)

FIGURE 7.11
FPGA-based prototype board. (a) Board photo and (b) Board configuration overview.

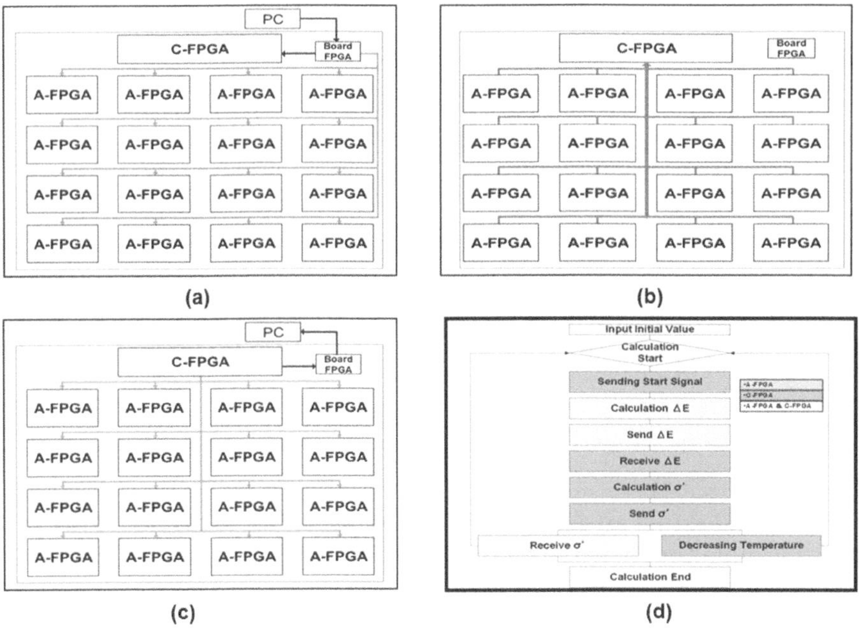

FIGURE 7.12
Overall calculation flow and calculation flowchart. (a) Step 1: Load data, (b) Step 2: Calculation part, (c) Step 3: Updating part, and (d) calculation flow chart.

the stored interaction values. When the calculation is completed, the calculation results are sent to the C-FPGA via serial communication through a 2-bit transmission line. These calculations are performed in parallel on multiple A-FPGAs.

Step 3 shown in (c) is the spin update part. The C-FPGA calculates the spin value using the spin value collected from each A-FPGA and external field data, temperature data, and random number data already recorded in the system. When the calculation of the spin value to be updated by the C-FPGA is completed, the spin value is sent to the spin transmitter/receiver dedicated The spin value is sent to each A-FPGA through the wiring (2 bits), and the A-FPGA receives the value. The spin value in the A-FPGA is then updated as the new spin value.

Figure 7.12d shows a more detailed flowchart of the calculation process based on (a)–(c). Steps 2 and 3 are repeated a sufficient number of times while the temperature is gradually lowered, and the calculation is terminated when the temperature reaches zero or the specified number of times. The solution to the problem can be obtained by reading the spin value at the end of the calculation.

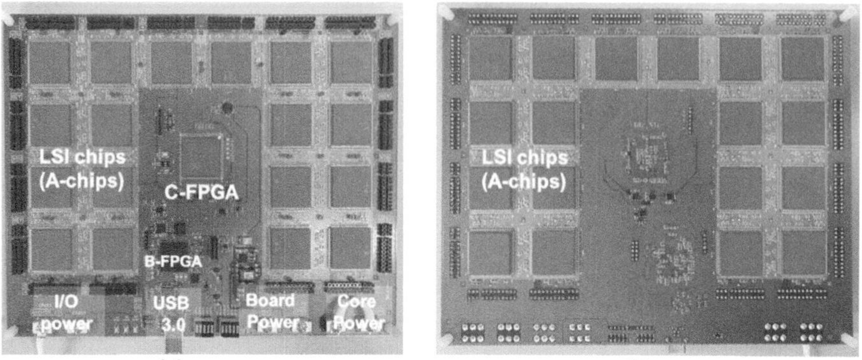

(a) (b)

FIGURE 7.13
Prototype board using ASIC LSI. (a) Board photo (front) and (b) Board photo (back).

Figure 7.13 shows an example implementation [7] of a 22-nm LSI with 512 spins mounted as shown in Figure 7.6 as an A-chip. In Figure 7.13, the number of spins implemented is eight times the number of spins mounted on one A-Chip. Since the A-chip implements 512 spins, it operates as a fully coupled Ising machine with 4,096 (4k) spins. It also has eight spin threads that can operate as if they were multiple times or multiple chips at once. This means that even if data is transferred in and out only once, eight calculations can be performed at once.

Figure 7.14 shows the measurement system and some characteristic values when solving the problem using this 4 k fully coupled Ising machine. In this system, the Achip was running at 10 MHz, and the power of the 36 Achip cores was 1.29 W. The total system power was 2.90 W. The results of solving a 4,096-vertex max-cut problem using this system are shown in Figure 7.15, comparing the proposed system (8 threads) with the results of emulating the same conditions on a CPU. The computation time was 3,341.5 s for the CPU and 98.3 s for the proposed system, which means that 8 solutions were obtained in 34.0 times faster computation time than the CPU. Here, the parameters of the proposed system, J_{ij}, are 4-bit while the CPU is 64-bit floating-point, but this was done under the condition where the calculations were performed as if they were 4 bits. The proposed system and the entire PC are compared as the same object. In addition, in the Core section, the power consumption of CPU only and 36 Achip only (Core+I/O) are compared. The proposed system has 2,344 times higher power performance ratio than CPU and can search eight solutions in parallel. 2,223 times higher power performance

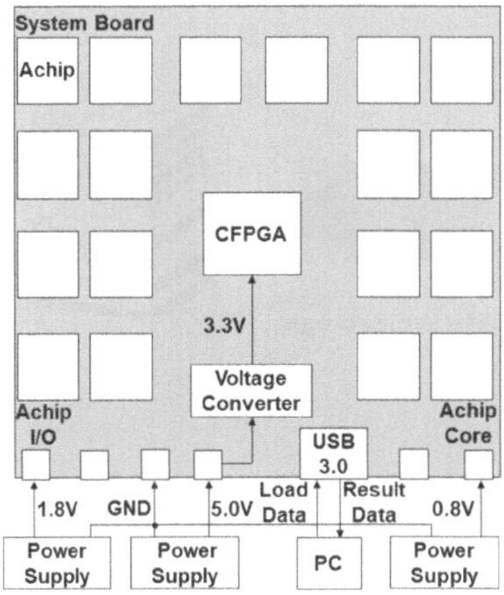

Entire System	A-chip36 + C-FPFA
Device (LSI chip / FPGA)	22nm CMOS / XC7S100-2FGGA676I
Op.Frequency [MHz]	10
Voltage [V] (Achip Core / Achip I/O / The others)	0.8 / 1.8 / 5.0
Op.Power [A] (Achip Core / Achip I/O / The others)	1.608 / 0.033 / 0.31
Op.Power [W] (Achip Core / Achip I/O / The others / Entire System)	1.29 / 0.059 / 1.55 / 2.90

FIGURE 7.14
4k-spin fully coupled Ising machine LSI system and some measured characteristics.

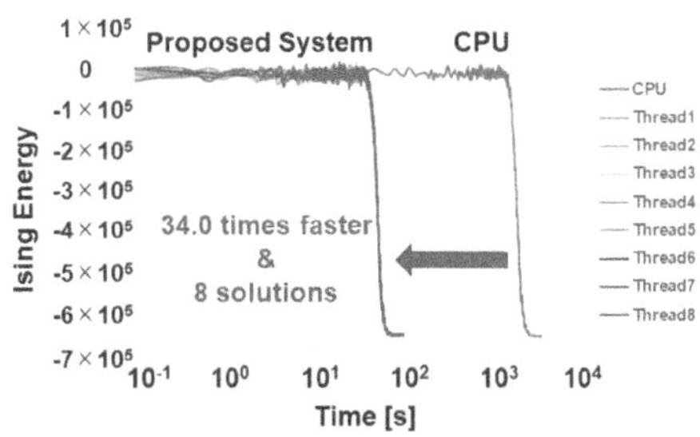

(a)

	CPU Simulation	This work
Op. Frequency [MHz]	3600	10
Op. Power [W] (Core part)	200 (85)	2.9 (1.3)
Time [s]	3341.5	98.3
Power consumption [J] (Core part)	657444 (279414)	285 (128)

(b)

FIGURE 7.15
Evaluation results of 4k spin fully coupled Ising machine LSI system. (a) Energy transition diagram for the max-cut problem and (b) performance comparison.

ratio is also confirmed in the comparison of the core part. Thus, one fully coupled Ising machine system can be created.

Figure 7.16 shows a schematic of the deployment of this method. Many spins can be implemented by using multiple boards or multiple chip stacks or by 3D mounting in a package.

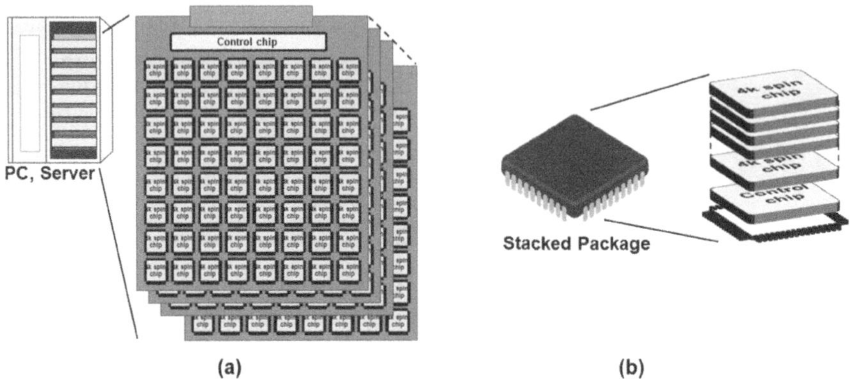

FIGURE 7.16
Scalable configuration development. (a) Multilayer board and (b) multilayer chip.

7.5 Advanced Fully Coupled Ising Machines

Fully coupled Ising machines are being studied intensively around the world. Here we present examples of efforts to improve the accuracy and speed up the computation.

For improved calculation accuracy, the Ising machine first expresses the problem to be solved in terms of J_{ij} pairs. Then, the solution is obtained by calculating the energy using these pairs. Therefore, the accuracy of J_{ij} is important for accurately representing the problem and obtaining the correct solution. We describe a method to solve this problem by scalability.

For computational speedup, we describe a method for simultaneous spin update that can be achieved by multivaluing the spins.

7.5.1 High Precision

While constructing an LSI with dedicated circuits enables high-speed computation and low power consumption, it has the limitation of being fixed to the initial specifications. The scalability described in the previous section allows the number of spins to be changed, but the bit width of the interaction (coupling coefficient) J_{ij} remains fixed. This constraint may make it difficult to solve certain types of problems. Here, we present a method [8,9] that simultaneously scales the interaction bit-width in addition to the spin number.

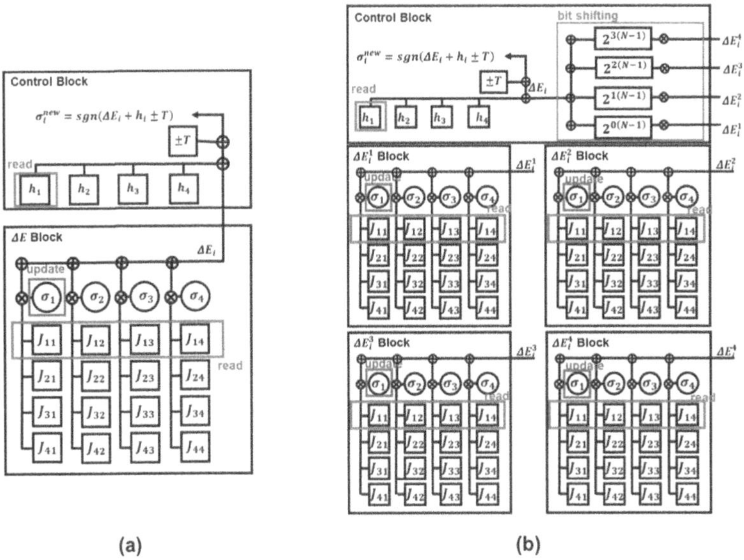

FIGURE 7.17
Scalable method for achieving high precision. (a) Basic structure and (b) high precision structure.

Figure 7.17a is a restatement of the method for updating σ_i. The i-th row of the interaction matrix is read and ΔE_i is calculated as the sum of its products with the corresponding spins. The sign of the value of ΔE_i plus $h_i \pm T$ was then used as the spin update value. Then, for scalability of the number of spins, the interaction was divided into n × n blocks, allowing ΔE_i to be split and calculated as separate chips. This method can also allow for expansion of the bit width of the interaction J_{ij} (high-precision structure).

In a high-precision structure, multiple ΔE blocks are provided, each of which performs different bit-level calculations. Figure 7.17b shows an example of four blocks: there are four A-chips corresponding to four blocks and one C-FPGA to integrate. In this case, when summing from the four LSI chips in the C-FPGA, bit shifting is performed, and then, the sum is taken. Let's put the sum of energy at this time as ΔE_i^h. Let's assume the following.

$$\Delta E_i^h = 2^0 \cdot \Delta E_{i1} + 2^3 \cdot \Delta E_{i2} + 2^6 \cdot \Delta E_{i3} + 2^9 \cdot \Delta E_{i4} \tag{7.10}$$

Here, ΔE_{ik} was the following for each A-chip.

$$\Delta E_{ik} = \sum_k J_{ik} \sigma_k \tag{7.11}$$

Therefore, equation (7.10) is equivalent to realizing J_{ij}^{h} integrated by bit shifting J_{ij} from each A-chip. For example, if the bit width of J_{ij} from each A-chip is 4 bits with a sign, then the integrated J_{ij}^{h} will be 13 bits with a sign since the bits are shifted and added up by 3 bits as in equation (7.10). This means that the signed 13-bit J_{ij}^{h} is put into four LSI chips with the same sign and the digits separated by 3 bits each, which means that a large bit width can be used to express the combination. In this way the J_{ij} bit width, which is the precision of the calculation, can be made scalable. Here, the capacity will remain the same for each original A-chip. This technique is as simple as scaling the number of spins. Therefore, the increase in a C-FPGA process is small. The only thing that needs to be considered is whether or not to apply bit shifting. Generalizing, this structure allows the C-FPGA to handle high bit-width ΔE_i^h by bit-shifting each ΔE_i^l from D A-chips. As shown, a signed$N$ bit interaction (J_{ij}^{l}) can be represented as a single signed$\{(N-1)(\ )D+1\}$ bit interaction (J_{ij}^{h}) using the bit-shift operation during addition. The relationship is expressed by the following equation.

$$J_{ij}^{h} = \sum_{l}^{D} J_{ij}^{l} \times 2^{(N-1)(l-1)} \tag{7.12}$$

The following equation by this is valid, and ΔE_i^h can be calculated by adding each ΔE^l.

$$\Delta E_i^h = \sum_{j} J_{ij}^{\text{high}} \sigma_j = \sum_{l}^{D} \Delta E_i^l \times 2^{(N-1)(l-1)} \tag{7.13}$$

Also, as can be understood from the previous description, this can utilize both high capacity structures (increasing σ_i the number of spins) and high precision structures (interacting J_{ij} bit-width extensions). These two structures compute ΔE_i at different stages. The ΔE_i^l in the high-precision structure in terms of the high-volume structure is expressed by the following equation

$$\Delta E_i^l = \sum_{k} \Delta E_{i,k}^l \tag{7.14}$$

Substituting equation (7.13) into equation (7.14), we obtain

$$\Delta E_i = \sum_{l}^{D} \left(\sum_{k} \Delta E_{i,k}^l \right) \times 2^{(N-1)(l-1)} \tag{7.15}$$

Figure 7.18 shows the development of a high-capacity and high-precision structure using the system with 36 512-spin 4-bit A-chips and one C-FPGA in Figure 7.13.

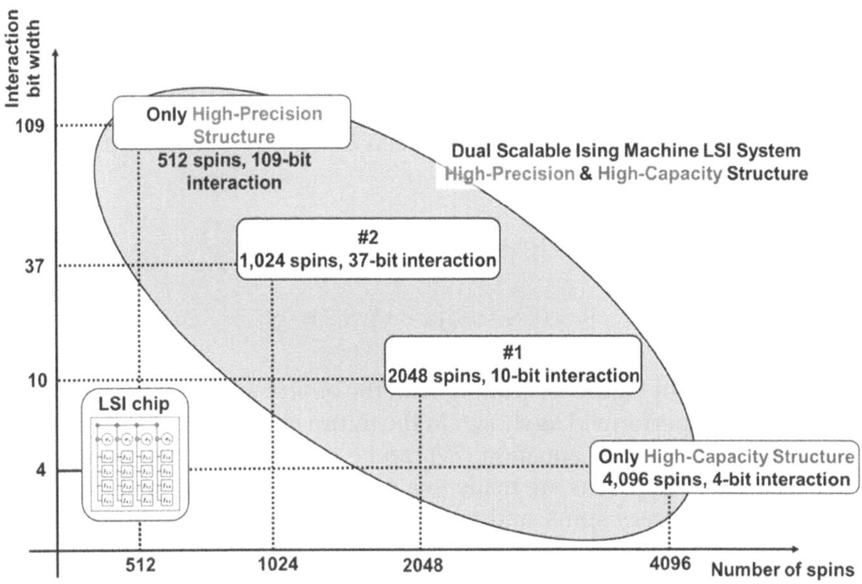

FIGURE 7.18
Dual Scalable Ising machine LSI System.

1. 2,048 spin-10-bit interaction
   The 30 LSI chips on the board are used in groups of 10, each group realizing 2,048 spins. Three sets of these were used to achieve a 10-bit bit width.
2. 1,024-spin 37-bit interaction
   For the 36 chips on the board, 1,024 spins were built using 3 LSI chips, and 12 pairs of these were used to achieve a bit width of 37 bits.

Furthermore, if all 36 LSI chips are used for bit-width expansion, the total bit width is 109 bits. This verification is limited to 36 LSI chips because they are already mounted on the board, but this technology can be deployed at any time for both accuracy and capacity axes using multiple LSI chips.

7.5.2 High Speed

Since simultaneous update of multiple spins is effective in speeding up the process, various developments have been reported as described in Chapter 6. In Ising LSI, the ground state is obtained by repeatedly updating the spin states. Therefore, a faster solution is expected by updating multiple spins simultaneously on the LSI. However, in many cases, simultaneous state updates for multiple spins do not converge as is.

Here, we show that due to multivalued spin states [10]. Instead of the binary spins of ±1 based on the Ising model, the method introduces multistate (MS) spins inspired by the Heisenberg model. $2K$ states are the following: 1) the spin state of the Ising model is the one in which the spin inversion is performed via the intermediate value between $2K$ states in the range±1 , unlike the conventional binary spins.

$$-1 = S_{-K}, S_{-K+1}, \cdots, S_{-1}, S_1, \cdots, S_{K-1}, S_K = 1$$

$$S_1 > 0,\ S_x < S_y\left(x < y\right), S_{-x} = -S_x$$

Then, as shown in Figure 7.19a along with the binary spin case, the spin state transitions are performed as shown in the figure when K is set to 4 (8 states). In this method, too, the equation (7.9) can be used as the spin update equation. The state transitions are made to a different sign when spin flipping is performed for binary spins, and to the same sign when spin flipping is not performed. This reduces the change in spin value per update. Therefore, even if simultaneous updates are performed, the change in the value of ΔE_i will be less than for sequential updates. This means that the basic algorithm of ΔE_i calculation and spin update can be implemented in parallel for the number of simultaneous updates. In other words, parallelization can be achieved with a simpler algorithm than the conventional method.

In the 1000-vertex Max-cut problem as an example, 1000 spins could be parallelized, and 64 spins could be parallelized in the traveling salesman problem with 256 spins in 16 cities.

In this example, LSIization does the following. The biggest difference from the conventional binary method is the calculation of the multiplication

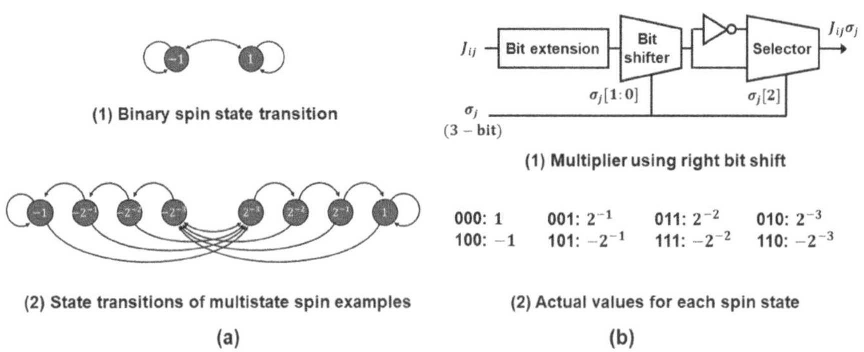

FIGURE 7.19
Multistate spin method for simultaneous updates. (a) State transitions of binary spin and multistate pin and (b) procedure for multiplication $J_{ij}\sigma_j$ using right bit shift.

$J_{i,j}\sigma_j$. Since spins have multiple states, using them directly in the calculation increases the size of the circuit compared to the binary spin method. Therefore, as shown in Figure 7.19b, the absolute value of each state is set to 2^{-n}. This allows the multiplication to be replaced by a right-bit shift, thus reducing the increase in circuit size and circuit delay. The block diagram of the multiplier using right-bit shift is shown in the same figure. Thus, the multiplier can be simplified.

7.6 Example of Seeking Solutions Using Scalable Fully Coupled Ising Machines

Let us show a few more results [6] from solving a easy-to-understand example of a combinatorial optimization problem using the fully coupled scalable Ising machine board shown in Figure 7.11.

Figure 7.20 shows the problem of coloring the 23 wards of Tokyo, Japan, so that they are not adjacent to each other in four different colors. Although the actual cost function and other details are omitted, this problem is a graph coloring problem where each side is represented as an adjacent graph. In Figure 7.20a, all the colors are not determined at first in the initial state (1), but after going through the intermediate state (2), the coloring of the 23 wards is finally completed in the state (3). The coloring of the 23 wards is completed in

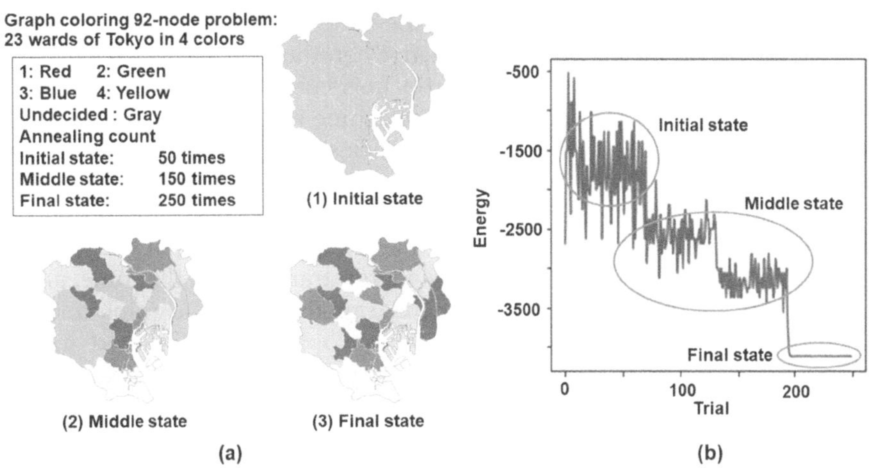

FIGURE 7.20
23 cities 4 color classification problem. (a) Problem solved and (b) energy transition diagram.

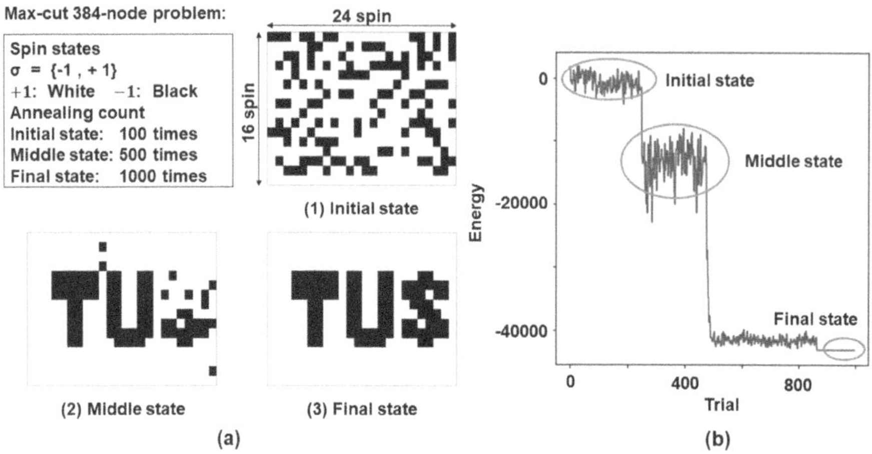

FIGURE 7.21
384 vertex max cut problem. (a) Problem solved and (b) energy transition diagram.

the final state (3). This result confirms its function as an annealing machine. Figure 7.20b shows the energy transition diagram for each state. The initial state has large oscillations and high energy, but as the number of steps increases, the energy oscillations and magnitude decrease, finally converging to a single state. The characteristic behavior is that the overall transition is discontinuous, rather than oscillating and decreasing gently as a whole.

Next, Figure 7.21 shows the max-cut problem. The max-cut problem was a graph represented by a vertex and a weighted edge, and the problem was to select a vertex of the graph so that the sum of the weights of the edges connected to the vertex was maximized when the vertex was divided into two parts. The current one is a 384-vertex max-cut problem, and Figure 7.21a shows the change in spin on the actual machine while the problem is being solved, with white showing spin +1 and black showing spin −1. (1) is the initial state with a shallow number of steps, and although the initial state of the spin is a messy state, it gradually converges to a character through state (2), and finally the spin state (3) converges to a state showing a character.

Figure 7.21b shows a diagram of the energy transition at this time, and it can be seen from the energy transition diagram that, as in the graph coloring problem, the energy is high when the initial spins are disordered, gradually decreases in energy, and finally converges to a single low-energy state when the letters are in the emergent stage The energy transition diagram shows that the energy is high when the first spin is turbulent and gradually decreases. The energy does not gradually decrease while oscillating, but there is a jump in the transition.

7.7 Summary

In this chapter, we focus on the versatility of the fully coupled Ising machine, which can handle combinatorial optimization problems in their original mathematical form. The main issue with the fully coupled type is that the number of connections between spins is one-to-many, which results in a huge number of couplings and places a heavy burden on the wiring density and circuit implementation inside and outside the chip. To address this issue, the case study presented in this chapter employs a configuration in which the spins are linearly arranged in one dimension and the interaction between spins (coupling coefficients) is developed on a two-dimensional array. Specifically, the arithmetic circuit corresponding to each spin and the interaction cell are connected in a vertical row to efficiently implement the fully coupled structure, thereby reducing the circuit size and allowing scalability in terms of the number of spins. Furthermore, by installing multiple rows of these 1D spin cells, it is possible to process multiple operations and multiple chip operations in parallel. For the two-dimensional array of interaction cells, a configuration that halves the number of cells by taking advantage of symmetry has also been introduced. For the spin update method, the "pseudo annealing method" is adopted, which can be easily implemented without analog circuits, and hardware realization is also considered to be easy. As for inter-chip coordination, a scalable configuration method is proposed in which multiple fully coupled Ising LSIs are connected using a very small number of interfaces, and each chip performs distributed processing of energy calculations while operating as a unified fully coupled system as a whole. As a more advanced method, a technique to improve the accuracy of interaction representation and a parallel update method to update multiple spins simultaneously by multistate the spin states are presented.

References

1. K. Someya, R. Ono and T. Kawahara, "Novel Ising Model Using Dimension-Control for High-Speed Solver for Ising Machines," *2016 14th IEEE International New Circuits and Systems Conference (NEWCAS)*, Vancouver, BC, Canada, 2016, pp. 1–4. https://doi.org/10.1109/NEWCAS.2016.7604797
2. R. Ono, K. Someya, and T. Kawahara, "A Novel Ising Model Processing Achieving All Interactions Only by Adjacent Spins for a High-Speed Solver for Versatile Ising Machines," *Microprocessors and Microsystems*, vol. 78, p. 103251, 2020. https://doi.org/10.1016/j.micpro.2020.103251

3. R. Iimura, S. Kitamura and T. Kawahara, "Annealing Processing Architecture of 28-nm CMOS Chip for Ising Model With 512 Fully Connected Spins," *IEEE Transactions on Circuits and Systems I: Regular Papers*, vol. 68, no. 12, pp. 5061–5071, 2021. https://doi.org/10.1109/TCSI.2021.3114422
4. S. Kirkpatrick, C. D. Gelatt, Jr. and M. P. Vecchi, "Optimization by Simulated Annealing," *Science*, vol. 220, pp. 671–680, 1983. https://doi.org/10.1126/science.220.4598.671
5. A. Endo, T. Megumi and T. Kawahara, "Fabrication and Evaluation of a 22nm 512 Spin Fully Coupled Annealing Processor for a 4k Spin Scalable Fully Coupled Annealing Processing System," *2024 IEEE 22nd World Symposium on Applied Machine Intelligence and Informatics (SAMI)*, Stará Lesná, Slovakia, *2024*, p. 000071, 2024. https://doi.org/10.1109/SAMI60510.2024.10432908
6. K. Yamamoto and T. Kawahara, "Scalable Fully Coupled Annealing Processing System and Multichip FPGA Implementation," *Microprocessors and Microsystems*, vol. 95, p. 104674, 2022. https://doi.org/10.1016/j.micpro.2022.104674
7. T. Megumi, A. Endo and T. Kawahara, "Scalable Fully-Coupled Annealing Processing System Implementing 4096 Spins Using 22nm CMOS LSI," *IEEE Access* , vol. 12, pp. 19711–19723, 2024. https://doi.org/10.1109/ACCESS.2024.3360034
8. D. Cui, T. Megumi, A. Endo and T. Kawahara, "Proposal and Validation of a Dual Scalable Annealing Processing System That Simultaneously Scales Capacity and Precision," *2024 International Conference on Microelectronics (ICM)*, Doha, Qatar, 2024, pp. 1–6. https://doi.org/10.1109/ICM63406.2024.10815759
9. D. Cui, T. Megumi, A. Endo and T. Kawahara, "Dual Scalable Annealing Processing System That Scales Number of Spins and Interaction Bit Width Simultaneously," *IEEE Access*, vol. 13, pp. 52592–52606, 2025. https://doi.org/10.1109/ACCESS.2025.3553542
10. K. Kanai and T. Kawahara, "Efficient Implementation of Parallel Annealing Method with Heisenberg Model," *2024 IEEE Asia Pacific Conference on Circuits and Systems (APCCAS)*, Taipei, Taiwan, 2024, pp. 55–59. https://doi.org/10.1109/APCCAS62602.2024.10808590

8

Semiconductor Memory and Computing

Let us return for a moment to the Neumann-type computer [1], which is widely used for the development of artificial intelligence LSI. A characteristic of the Neumann-type information processing system is that it uses an architecture that separates computing and memory. In other words, in the hardware execution of information processing operations, data are passed back and forth between the storage device (memory) and the processing unit (CPU (central processing unit) or general processor), from memory to the processing unit and vice versa. The original data is read from memory, the necessary operations are performed, and the results are sent back to memory as data. The speed and smoothness of this data transfer are important for high-speed and efficient signal processing. In artificial intelligence processing, high-performance CPUs and GPUs have been necessary in the past to process large amounts of data. However, power consumption is increasing, and issues such as cooling are becoming more significant. In addition, memory capacity becomes a bottleneck when handling large data sets. This could reduce processing speed. In addition, higher performance is required for the input/output devices themselves. Information from sensors, etc., needs to be processed directly on site.

The major challenge here is that data are itself transferred back and forth from the memory to the arithmetic unit and vice versa. This data transfer consumes a lot of power and is slow. The operating speed of the memory itself is also slow relative to the arithmetic unit.

Therefore, many efforts have been made to solve this problem and improve the performance of information processing and reduce power consumption. Among these, the integration of memory and arithmetic devices may be a solution. However, in normal computing, this often limits the versatility of information processing.

On the other hand, in the operation of a neural network, which is the subject of this book and is an artificial intelligence process, as has been explained so far, a number of fixed processes are performed, so to speak, in which the weights stored in memory and the results of the previous stage's operations are used to perform a sum-of-products operation. Moreover, by changing the weights, the processing can be made more versatile. Therefore, by utilizing this characteristic, it is possible to transfer data between the memory and the arithmetic unit at higher speeds and lower power without losing the versatility of the processing. Furthermore, it is suitable for eliminating the interface portion used for data exchange and incorporating the arithmetic mechanism directly

DOI: 10.1201/9781003599067-8

into the memory. This will also enable low power consumption in combination with the characteristics of nonvolatile memory, which is under development.

In this chapter, we will learn about the role of memory in conventional signal processing, followed by a look at the characteristics of the main types of memory used in signal processing in Chapter 9. Based on the above, in Chapter 10, we will learn about near-memory computing, in which the arithmetic unit and memory are placed close to each other, and in-memory computing, in which the arithmetic unit and memory are integrated.

8.1 Computing and Memory

The role of memory in computing is to hold information and supply information to the processor (CPU). For example, when extracting necessary information from voice data, calculations are required. The processor does this, but in order to perform the calculation, the target speech data must first be retrieved from the memory that stores and holds it. The information processing proceeds by retrieving the data, performing the calculations, and storing the results back into memory.

In most modern computers, the architecture defines simple instructions such as read, write, add, multiply, etc., and these instructions are commonly used in each application, suitable for the target application. To improve performance, these instructions must be executed as fast as possible.

The hardware configuration [1] consists of a control unit, an arithmetic and logic unit (ALU), a storage unit, an input device, and an output device, as shown in Figure 8.1. The storage unit consists of a main memory

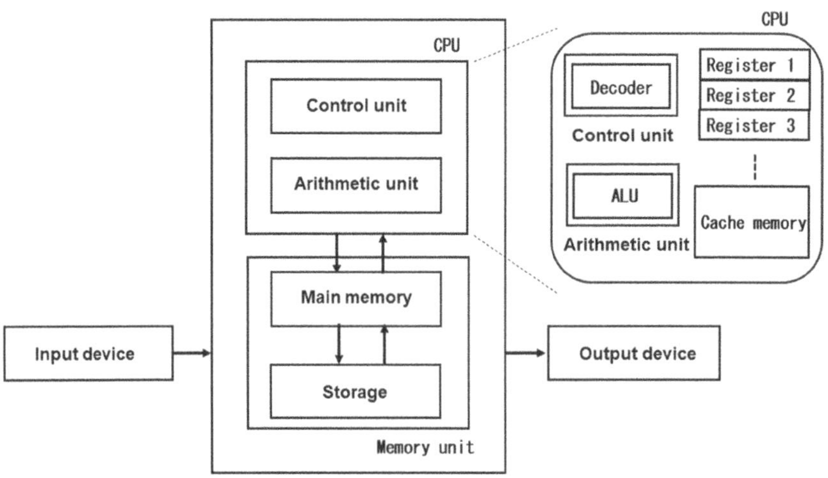

FIGURE 8.1
Computer configuration.

(main memory), a memory unit, an input device, and an output device. The storage unit is classified into main memory (main storage/RAM), storage (external storage/auxiliary storage), etc., but it also includes the CPU's internal cache memory. The control unit (control unit) decodes instructions, manages data flow, and communicates with input/output devices, while the arithmetic unit (ALU: Arithmetic and Logic Unit) processes data and performs calculations. The performance of an arithmetic unit is evaluated in terms of processing speed and the number of bits of data that can be processed simultaneously (32-bit, 64-bit, etc.). The power consumption of a CPU ranges from several watts to several tens of watts. An input device is a device used to provide data, information, commands, etc. to a computer, and generally refers to a device that is operated by a human being to provide input. Examples include keyboards, mice, touch panels, and microphones. Output devices are devices that physically present the computer's internal information and results to the outside world in a form that is recognizable to humans. Examples include a display-mo or projector, printer or plotter, speaker or earphone.

In the operation of a computer, main memory is an important element that temporarily holds data and programs and provides the information necessary for processing; the data and instructions necessary for the CPU to execute programs are stored in memory, to which data access is made as efficiently as possible. The CPU uses these to perform processing. Storage is a device used to store data and programs for long periods of time. HDDs and SSDs are commonly used. SSDs have much faster reading speeds than HDDs, sometimes reaching several GB/s. HDDs have larger capacities but slower speeds, around several hundred MB/s. SSDs are used to store data and programs for long periods of time. In program execution, program instructions are also stored in memory; the CPU is equipped with a program counter to retrieve the next instruction to be executed from memory. This enables sequential execution of programs.

There are several issues with the memory system as a whole. First, CPUs perform processing at high speeds, but main memory is relatively slow in comparison, and storage is even slower. In addition, there is a trade-off between memory access time (the time it takes to access data) and the storage capacity of that memory. In other words, memory with fast access is generally small in capacity, while storage (external or auxiliary storage) is large in capacity but slow in access. In addition, when a program is executed, some data is used frequently, but some data is used only temporarily.

8.2 Memory Hierarchy in Computing

Because of these issues, multiple types of memory are used in general computing as shown in Figure 8.2, including PCs, portable information terminals such as cell phones, and information appliances such as digital TVs,

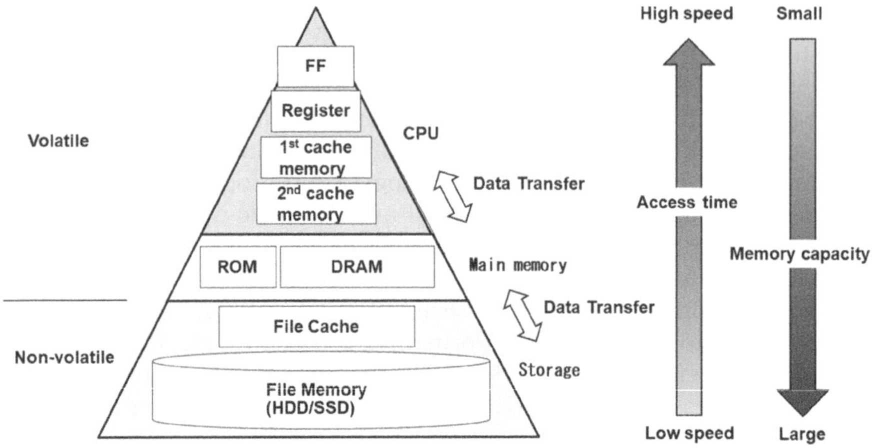

FIGURE 8.2
Memory hierarchy in computing.

depending on their role [2]. For the CPU, there is main memory and storage, and cache memory is also placed inside the CPU. Here, nonvolatile memory is used for storage and program storage, while volatile memory is used for main memory and cache memory near the CPU. As described below, non-volatile memory is memory that continues to retain memory even when the power is turned off, while volatile memory is memory that disappears when the power is turned off.

a. Cache memory

The cache memory itself also has a hierarchical structure for high-speed processing: the primary cache memory (L1: Leve 1 Cache Memory, 1st cache) is an extremely fast memory located inside the CPU that stores the data and instructions most frequently used by the CPU core. This allows the CPU to quickly retrieve necessary data without accessing the main memory (RAM). Typically, the L1 cache access time is less than one cycle of that CPU's operating frequency. The secondary cache memory (L2: Level 2 Cache Memory, 2nd cache) has a larger capacity than the primary cache and is not accessed as often as the primary cache, but it holds a large amount of data. If the L1 If the L1 cache misses, the L2 cache is consulted next. This allows necessary data and instructions to be efficiently retrieved during program execution, thereby increasing processing speed. Access times typically take several cycles. The capacity is about 512 KB–4 MB. Some systems have even deeper caches (tertiary or higher). These cache memories use SRAM, described below, which can be accessed at high speed. The CPU also has an FF, which is used to temporarily hold data. In cache memory, it is important that the data required by the CPU exists in the cache memory, and the ratio of the data required for the operation that exists in the cache memory is called the cache hit rate. A high cache hit rate reduces the CPU's access to slow RAM and improves overall processing speed. If a cache miss occurs, the CPU must access main memory, resulting in a slower processing speed.

b. Main memory

The main memory is located between the processor and the storage device. It is a high-speed memory that temporarily holds programs and data and can be accessed and processed by the CPU directly as well. The access time is several tens of ns. However, it also has a mode to transfer data collectively at high speed. It holds the program being executed and data being processed so that the CPU can quickly retrieve the data. The main memory is volatile, and data are lost when the power is turned off. Main memory is also known as RAM (Random Access Memory). Capacity generally ranges from a few GB to a dozen GB. DRAM, described below, or SRAM is used in embedded devices. Read-only ROM that stores frequently used instructions, etc., may be included in this category.

c. Storage memory

Storage memory is a device for long-term storage of data and is primarily nonvolatile memory. It stores the operating system, applications, and user data of mobile devices. As storage memory, hard disk drive (HDD) is a storage device using magnetic disks and has a large capacity but slow access speeds of only a few milliseconds (ms). Solid state drive (SSD) is a storage using flash memory, described below, and can be accessed in tens of microseconds (μs) but costs more than an HDD. This hierarchical structure of memory in computing is intended to provide efficient data access and processing speed. Main memory is relatively fast but has limited storage capacity, including in terms of cost. Storage devices, on the other hand, have large storage capacity but slow access speeds. This hierarchical structure aims to achieve both high-speed processing and large data retention. In addition, since the CPU often accesses the same data and instructions many times, the use of a high-speed cache memory can efficiently hold repeatedly accessed data and reduce the amount of data transferred between the CPU and main memory.

However, there is a limit to the performance improvement of this memory hierarchical structure, because the data transfer part is inevitably the performance-limiting factor. Even if the CPU operates at GHz, the input/output section with the memory is one digit slower. Furthermore, despite being slow, the power consumption associated with signal propagation is high. The parasitic capacitance and parasitic resistance of the signal lines connecting the two chips are three orders of magnitude higher than those in the chips themselves, and as we enter the IoT era and process large amounts of data, solving this problem is essential. This is especially true for processing at the edge, where power capacitance is limited.

In a data-centric processing system, the idea of the countermeasure itself is simple. It is simply to place the arithmetic circuits and the memory in which the program and data are stored as close as possible. This is achieved by near-memory computing. In-memory computing is a configuration in which the computing circuits and memory are integrated and processing is executed in the memory, thereby further reducing the amount of data movement. In artificial intelligence processing, in-memory computing is effective in improving performance because sum-of-products operations are performed using weights stored in memory and the results of previous operations.

However, the degree of freedom is lower than in the case where arithmetic circuits and memory are individually enhanced and connected at will. By solving this limitation by specializing in artificial intelligence processing, the advantages in terms of power efficiency and implementation cost become significant.

8.3 Classification of Semiconductor Memory

Let us look at the types of semiconductor memory. Advances in semiconductor technology have greatly advanced computer memory. The main performance and characteristics of semiconductor memory that should be considered are as follows. Access speed is important in terms of computing performance. Although it is the speed of reading and writing to memory, the data transfer speed is often important. In addition, if the access speed of memory is fast, as a hardware system construction, the electrical properties of the board on which this is implemented will have a large impact. The next question is whether the memory is nonvolatile or nonvolatile, which is also the main classification of memory. Nonvolatile memory should be selected if the data needs to be retained even when power is turned off. This is not only for storage, but also for low-power operations, and the use of this nonvolatile memory as main memory is being considered. However, this nonvolatile memory has a limit to the number of times data can be written to it successfully, and this must be taken into consideration. In addition, memory with high access speeds and memory with a large number of read/write cycles tends to be expensive. It is necessary to select the appropriate memory for each application to minimize cost increases.

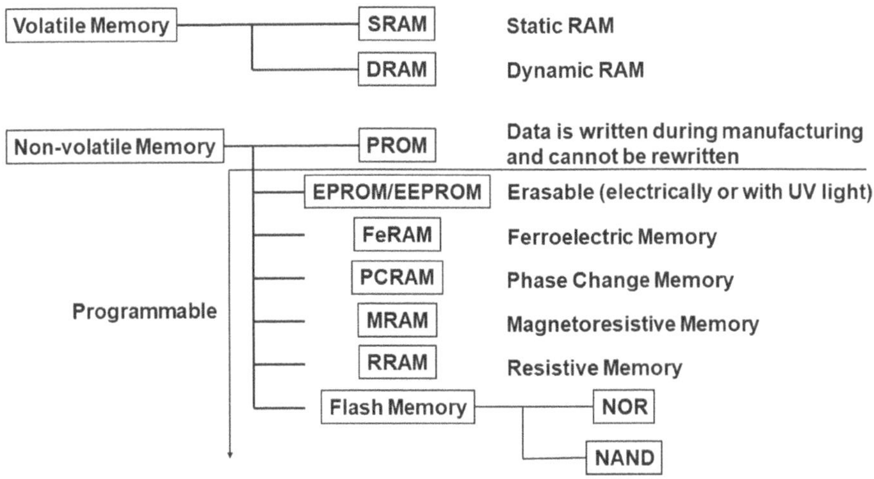

FIGURE 8.3
Classification of semiconductor memory.

Semiconductor memory [3,4], which is mainly used, is classified as shown in Figure 8.3. There are two major types of semiconductor memory: volatile memory and nonvolatile memory.

Volatile memory is used as a temporary work area for data handling in information processing because the memory is erased when the power is turned off. Volatile memory is usually divided into two types: DRAM and SRAM. SRAM is a flip-flop with two CMOS inverters for the memory cells and is capable of high-speed data access. This is useful when high-speed data processing is required in signal processing. DRAM is also used in small-capacity cache memories and register files, etc. DRAM has a larger capacity than SRAM because the memory cell consists of a single transistor and a single capacitor. It is used to efficiently handle large amounts of data in signal processing flows. However, it operates at a slower speed than SRAM, and the capacitor must be constantly refreshed to retain information.

Nonvolatile memory is utilized as storage (auxiliary storage device) because of its property of retaining data even when the power is turned off. It has been used for long-term storage of operating systems, applications, and user data. Here, what is generally called nonvolatile memory has a limited number of rewrites due to its rewrite principle. However, image data, music data, or programs are not frequently rewritten in response to CPU operations, and nonvolatile memory has been put to practical use while taking into consideration the limitation on the number of rewrites. The number of read operations is the same as that of volatile memory, except for some systems. Nonvolatile semiconductor memories include PROMs, EPROMs, EEPROMs, and new memories such as FeRAM (ferroelectric memory), PCRAM (phase change memory), MRAM (magnetoresistive memory), RRAM (resistance change memory), and flash memory, as shown in Figure 8.3. PROMs can be written only once, EPROMs can be erased with ultraviolet light and written electrically, and EEPROMs can be erased electrically. Some new memory and flash memory will be described later.

Nonvolatile memory is also being used in computing to take advantage of the fact that data are not lost even if the power is turned off. For example, if nonvolatile memory can function as main memory, system startup and shutdown will be quicker. In addition, data can be retained even when the power is turned off, eliminating the need to keep the power on all the time and reducing overall power consumption. Thus, there is also an attempt to combine nonvolatile memory with volatile memory (such as DRAM) to achieve both high-speed data access and nonvolatility. This can reduce power consumption while optimizing overall system performance.

Here, we first summarize the principle of nonvolatility of nonvolatile memory. This principle is important to understand when applying AI processing.

8.4 Nonvolatility Principle of Nonvolatile Memory

Figure 8.4 schematically illustrates the principle of how nonvolatile memory can retain memory in a nonvolatile manner. (a) is called FeRAM (ferroelectric memory), in which a large electric field is applied to change the location of atoms in the crystal, although the change is very small. The two different

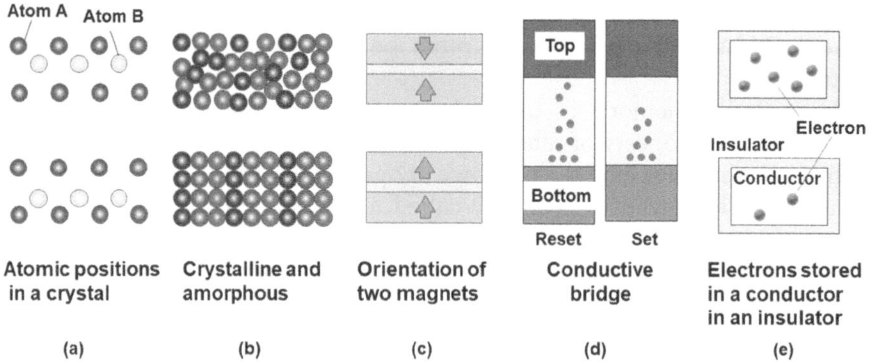

FIGURE 8.4
Principles behind nonvolatile memory.

locations of the atoms result in different polarization states, which are used as bit information. The location of the atoms hardly changes until a large electric field is applied again. However, each time a rewrite is performed, the location of the atoms is changed, and although the change is minute, the crystal breaks down after a certain number of rewrites. For this reason, there is a limit to the number of rewrites. (b) is to create two states by first applying heat that reaches several hundred degrees Celsius to the material that makes up the memory cell and then cooling it rapidly or by keeping it at a constant temperature (lower than the initial temperature, but several hundred degrees Celsius) for a while. It is called phase-change memory. When quenched, it becomes a material called amorphous (amorphous), and when kept at a constant temperature for a while, it becomes a crystallized material. Both have different electrical resistances, which correspond to information. At normal temperatures, these states do not change for long. This is called phase-change memory. However, this one is also baked or cooled during rewriting, and it will break down after a certain number of times. In MRAM (magnetoresistive RAM) (c), the sandwich structure (TMR element) consists of a thin insulating film sandwiched between two magnetic films, and information is written by changing the relative orientation of the two magnets by means of a magnetic field or STT (see below). The two magnetic films have low electrical resistance when the magnets are oriented in the same direction, and high resistance when they are oriented in opposite directions (tunnel magnetoresistance effect (TMR effect)). The direction of the magnet changes as the internal state of the electrons in the magnetic film changes. Therefore, in this MRAM, changing the overall orientation of the magnet has only a slight impact, but the number of rewrites of the MRAM is said to reach the number of times that volatile memory can be realized. (d) is called RRAM (Resistive RAM), in which an insulator exists between two electrodes and the composition of this insulator changes. There are various methods, but for example, this insulator is a solid solution in which metal atoms dissolve. Therefore, by applying a voltage, the metal from the electrodes is taken up, and a path is formed as a conductor. The formation of this pathway is caused by a large voltage. Small voltages at readout do not change this state, and the two states, insulating or conducting, correspond to the information. Because this memory is simple in structure, it is suitable for large capacities, although there is a limit to the number of rewrites. Finally, (e) is called flash memory and is widely used. Information is rewritten by using the tunnel effect in a conductor inside a slightly thick insulator to metaphorically knock electrons in or out. For readout, the presence or absence of electrons in the conductor causes the electric field seen from outside the insulator to propagate differently, and this is detected and read out by the small electric field during normal transistor operation. This method is suitable for large capacitance because of its simple structure. However, the insulating film deteriorates with each repeated rewrite, and after a certain number of rewrites, a hole opens in the insulating film.

8.5 Basic Circuit Configuration of Semiconductor Memory

In a system, the signal lines between the arithmetic unit (CPU) and memory are electrically connected by a bus. First, an address is sent from the CPU to the memory through the address bus. This address is interpreted by the address decoder in the memory LSI, which identifies the memory address and selects the corresponding memory cell. Next, data from these memory cells is output on the data bus, and the CPU reads this data. This data are used to perform calculations inside the CPU. The basic operation of the memory in this configuration is shown below.

The operation of a semiconductor memory is to read and write data to memory cells that store memory. The memory cell at a specific address is accessed, and data are written to it. Generally, data is binary, either "1" or "0". In the read operation, the memory cell at the address where the data is located is accessed, and the data are read out. Figure 8.5a shows the basic structure of a semiconductor memory for this purpose [3]. First, in order to achieve a large capacity, memory cells are laid out in two dimensions (matrix) to form a memory cell array. Then, a particular memory cell in this array can be designated by a word line, which is a selection line in the row direction, and a bit line, which is a selection line in the column direction (and can also be a read line and a write line). These are driven by the respective drivers so that they are selected by the row and column decoders based on the address signals *Ai* and *Aj* in Figure (a). With this configuration, if there are n *Ai* and m *Aj* address signals, 2^n word line (*WL)* signals are generated by the row decoder and 2^m bit line (*BL*) signals are generated by the column decoder. At the intersection of these word lines and bit lines, memory cells can be placed and selected, so that the memory cell array consists of 2^{n+m} memory cells. For example, a total of 20 *Ai* and *Aj* address signals can select 1M memory cells, and a total of 30 address signals can select 1G memory cells.

The bit lines are connected to a sense amplifier and a write circuit, whose output is *Do.* In addition, there is a control circuit that controls these circuits, represented here by *Cm,* and the signals necessary for operation are input. Data readout is divided into row readout and data transfer (column readout). In the row readout shown in Figure 8.5b, one word line is selected and a number of memory cells connected to it are read out simultaneously, and signals appear on the same number of bit lines. This signal is amplified by a sense amplifier. After amplification, the signals are stored in the same sense amplifier as a latch or in a latch aligned with the bit lines. A large amount of data is read out simultaneously inside the chip, but at a slower speed due to the driving of the word lines and the amplification of the bit line signals. It is important to note that the signal readout and amplification operations for the multiple bit lines are performed in parallel. We can attempt to add signal processing here. In other words, if a signal processing function can be provided for each of the multiple bit lines, the desired signal processing

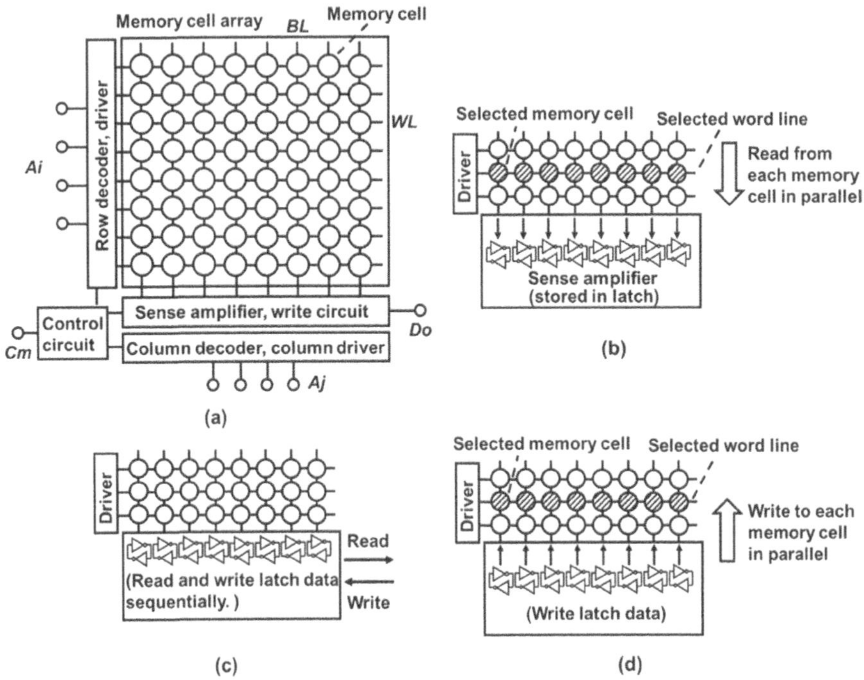

FIGURE 8.5
Array structure of memory LSI and read/write operations. (a) Basic structure of semiconductor memory, (b) row read, (c) data transfer, and (d) write.

can be performed simultaneously with the signal amplification operations of the multiple bit lines that are performed in parallel. Alternatively, a circuit that directly uses the signals of these multiple bit lines for signal processing can be placed side-by-side at the sense amplifier location. In this case, in the actual chip, it is important to match the spacing of the bit lines with the dimensions of such a circuit. The emphasis in this operation is on the large number of these multiple bit lines; the range is from 64 to about 1k lines. In other words, when "doing the desired signal processing," it is this number that is being processed in parallel.

In the data transfer (column read) shown in Figure 8.5c, data is transferred from the latch to the outside in synchronization with the clock. This transfer is high-speed. Generally, the amplification operation by selecting the word line takes about 10 ns, but the column readout can be synchronized to a high-speed clock of about 1 ns.

Writing data is the reverse of reading. First, data are transferred from outside the chip to a latch or sense amplifier connected to a bit line for one word line, as shown in Figure 8.5d. Next, when the word line is selected and the memory cell is selected, the information stored in the corresponding latch or sense amplifier is sent to the memory cell, and the data are written to the memory cell.

Normally, the structure shown in Figure 8.5a is mounted on a single memory LSI chip, but in what is called a bank structure, multiple units of this structure are prepared on a single chip. In the bank structure, multiple of these units are provided on a single chip. Each of these units can select word lines and amplify memory cell signals with bit lines as one bank. In this structure, since word line selection is possible for each bank, information from multiple memory cells connected to multiple word lines on a single chip is simultaneously read out to the respective bit lines, and the data are read out to the sense amplifiers and latches connected to the bit lines. In the bank structure, data are continuously transferred out of the chip from latches provided in multiple banks. Furthermore, in parallel with the transfer of data from the latches in one bank, a row readout can be performed in a different bank. Combinations of these operations and signal processing are also being considered.

The above configuration is almost the same for many types of memory, as shown in Figure 8.3. However, the performance of a memory cell differs depending on its configuration, including its read speed, rewrite speed, how many times it can be rewritten, and how long the stored information can be retained. Understanding the characteristics of memory cells and considering how to select and use the most appropriate memory is essential for advanced signal processing.

Thus, as a memory cell array, read/write operations are performed on memory cells laid out in two dimensions. On the other hand, as shown in Figure 1.4 in Chapter 1 and Figure 4.9 in Chapter 4, as a configuration for AI processing, PEs were laid out in two dimensions. Furthermore, taking inference as an example, the basic operation of the neural network was to take the product of the weight data determined by training and the input. Here, this weight data is stored in memory. Therefore, if the weight data is stored in memory and the product of the weight data and the input is taken when this memory is read out, the memory can be read out and the operation can be performed at once. Of course, the weight data is different from the data in a single memory cell. However, this idea is the principle of in-memory computing, which will be discussed in Chapter 10, because it is also consistent with the basic operation of neural networks.

In addition, the operation in (b) is a read/write operation to the sense amplifier row, which is different from the parallel operation from the sense amplifier to the memory cells in (a) and (c). Therefore, it is conceivable to incorporate neural network operations here. This type of operation can also be considered in the category of computing-in-memory.

In this way, high performance and low power can be achieved by utilizing the array structure of memory in a memory chip or memory block to perform the basic operations of a neural network.

8.6 Summary

This chapter provided an overview of the role of semiconductor memory in computing. Memory plays an essential role in storing data and increasing processing speed in computing. In general, memory in computing has a hierarchical structure, starting from the side close to the CPU: cache memory, main memory (main memory), and storage memory. This is a result of limitations in the performance and capacity of available memory. Memory is broadly classified into volatile memory (DRAM, SRAM) and nonvolatile memory (flash memory, MRAM, etc.). Each nonvolatile memory holds information based on a different principle. Furthermore, the basic circuit configuration and operation of semiconductor memory are shown. Memory cells are laid out in two dimensions to form a memory cell array, and specific memory cells are designated by word lines, which are selection lines in the line direction, and bit lines, which are selection lines in the column direction, for reading and writing.

References

1. J. L. Hennessy and D. A. Patterson, *Computer Architecture: A Quantitative Approach*, 6th ed. Morgan Kaufmann, 2022.
2. T. Kawahara and H. Mizuno, *Green Computing with Emerging Memory*, Springer Nature, 2013.
3. K. Itoh, *VLSI Memory Chip Design*, Springer, 2001.
4. S. Yu, *Semiconductor Memory Devices and Circuits*, 1st ed. CRC Press, 2022.

9

Main Semiconductor Memory Features

Although the composition of memory cell arrays is almost the same, the performance of semiconductor memory differs depending on the type of memory cell used. To consider the application of semiconductor memory to artificial intelligence integrated circuits, it is important to understand the structure and properties of each memory cell, its access speed, power consumption, durability against rewrite cycles, and reliability of the stored information. In this section, examples of typical semiconductor memories are discussed.

9.1 Volatile Memory

Volatile memory is memory that is erased when the power is turned off. It is used as a temporary work area when handling data. There are usually two types of volatile memory: SRAM and DRAM, with SRAM being fast but low capacity and DRAM being relatively slow but high capacity. DRAMs need to be rewritten at regular intervals to refresh their memory contents.

9.1.1 SRAM

Static random access memory (SRAM) [1] usually consists of two crossed inverters (M1 and M3, M2 and M4) as shown in Figure 9.1a. These inverters constitute a flip-flop circuit and hold one bit of information. In addition, two access transistors (M5 and M6) are present to control access. When an address is specified, the corresponding word line is selected, and the access transistors are turned on. This allows data to be read or written. When the access transistor is turned off, the stored 1-bit information is retained in the flip-flop circuit. This characteristic eliminates the need for refresh and rewrite operations, as in the DRAM described next. In addition, high-speed operation is possible, and random access is possible. As mentioned above, it is used in cache memory and register files.

SRAM is structurally compatible with CMOS process technology. Figure 9.1b shows the layout diagram, which consists only of CMOS transistors. Therefore, it can be composed of only CMOS transistors and wiring on a silicon wafer. Therefore, the SRAM can be easily integrated into

DOI: 10.1201/9781003599067-9

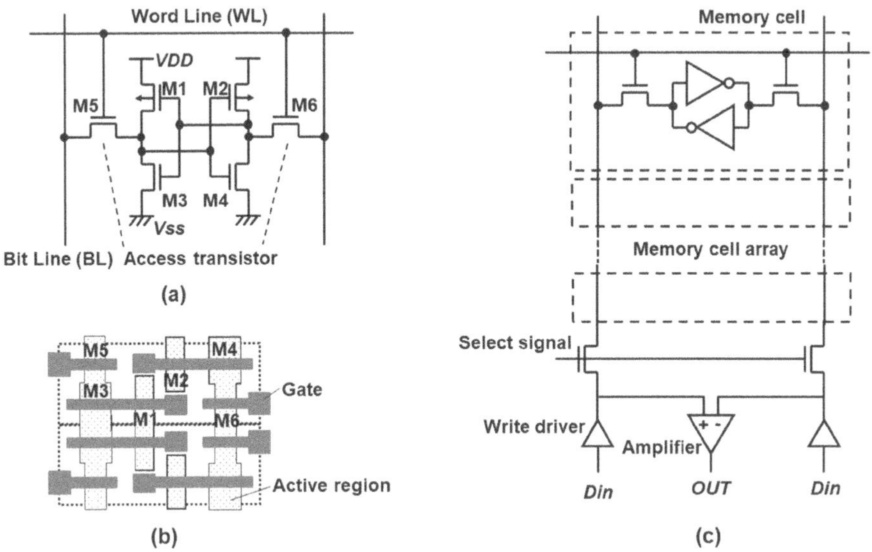

FIGURE 9.1
SRAM memory cell and basic array circuit. (a) Memory cell circuit, (b) memory cell layout, and (c) SRAM data path.

the arithmetic circuit block. In addition, it can operate at low voltage and has high power efficiency, making it suitable for mobile devices and power-saving applications. In terms of area, if F is the minimum line width for the microfabrication technology used in manufacturing, it is typically 120–150F [2]. This is large compared to other memories, but it is compatible with CMOS process technology, so it is easy to benefit from miniaturization. With miniaturization, transistor size is reduced, increasing the level of integration and enabling the production of high-performance SRAM. However, as miniaturization progresses, issues such as reduced stability and increased leakage current also arise. To address these issues, many design techniques have been developed, including transistor size optimization and word line and bit line drive methods.

Figure 9.1c shows the data path of the SRAM. It consists of a memory cell array, a sense amplifier, and a write driver. In the read operation, the accessed cell generates a differential voltage on the bit line pair. This is amplified by the sense amplifier and becomes the output *OUT*. In a write operation, the write driver sends write data *Din* to the selected bit line.

The SRAM memory cell operation is again summarized in Figure 9.2. In the data retention shown in Figure 9.2a, the access transistor is off, the voltage does not change, and the stored data is held stably. In this state, data are persistently retained as long as the power supply is maintained. During the write operation shown in Figure 9.2b, the word line is selected, the access transistor is turned on, and data are written into the cell via the bit line by the

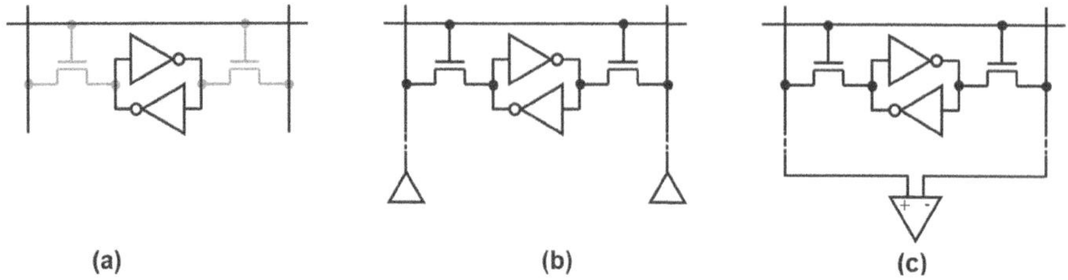

FIGURE 9.2
Basic operation of SRAM. (a) Data retention, (b) write, and (c) read.

write driver. During read, shown in Figure 9.2c, the cell is precharged before reading. When the word line is selected, differential signals appear in the two bit lines, and the sense amplifier reads the voltage difference.

Here, SRAMs have the advantage of high-speed data read and write. Dual-port and multiport SRAMs take this to the next level by having multiple ports that allow different operations to be performed simultaneously. This allows simultaneous read and write operations to a single memory cell, which is especially important for applications that require real-time processing. It also allows multiple processors to access the memory at the same time, improving overall system performance. Figure 9.3a shows an example of an 8T SRAM, a dual-port SRAM that uses eight transistors and has independent read and write ports. This structure optimizes the read and write operations independently. In this figure, RWL is the read word line, WWL is the write word line, RBL is the read bit line, and WBL,/WBL is the write bit line pair. This 8T SRAM has the following advantages along with high speed. First, the 8T SRAM does not affect the cell state when reading data. This increases the signal-to-noise ratio and improves stability during read operations. This means stable operation at low voltages, which enables high-speed operation while reducing power consumption. This is especially important for battery-powered devices. Furthermore, it has been shown experimentally to be highly resistant to soft errors caused by radiation, as described below, and other external factors. In addition, 10T SRAMs are also being developed for high-speed and low-voltage operation.

Soft errors are temporary malfunctions caused by external radiation or other factors. They are called soft errors because they do not damage the hardware itself and can usually be corrected by rebooting or rewriting data. As shown in Figure 9.3b, it is caused by both alpha rays and cosmic rays. Alpha rays are emitted from trace amounts of radioactive materials contained in the packaging materials of semiconductor devices. When alpha rays strike the silicon substrate of the semiconductor, an electrical charge is generated, which causes the data in the memory cell to change. Alpha particles do not penetrate deeply into the silicon, but they affect the package directly from the surface, making them difficult to defend against.

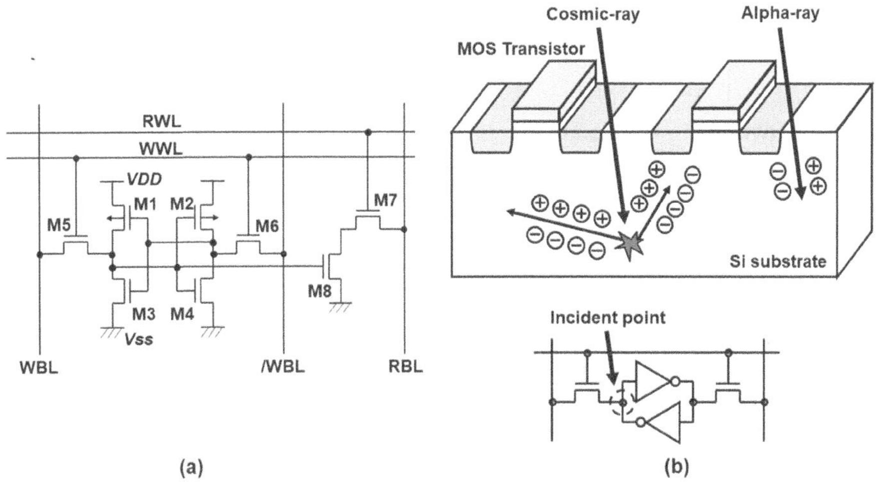

FIGURE 9.3
8T cell and soft error. (a) 8T dual-port SRAM cell and (b) soft error.

Cosmic rays are high-energy particles that enter the Earth's atmosphere. Secondary particles produced when cosmic rays collide with atoms in the atmosphere affect them. When these collide with silicon substrates, they produce an order of magnitude more charge than that produced by alpha rays, causing soft errors. These issues also apply to DRAM.

9.1.2 DRAM

Dynamic random access memory (DRAM) [2] generally consists of one capacitor for data retention and one access transistor, as shown in Figure 9.4a. The amount of charge held in the capacitor expresses the information of 1s and 0s. The simplicity of the structure allows for large capacitance. However, the data stored in the capacitor is lost over time due to leakage current through the access transistor and other components, making it impossible to read the data. For this reason, a refresh operation is performed to maintain the stored data in the DRAM. In the refresh operation, data are read from a selected DRAM cell, amplified, and written back to that DRAM cell. In addition, manufacturing costs increase because an additional process to the CMOS process technology is required to create the capacitor. Furthermore, as shown in the cross-sectional structure example, DRAM memory cell capacitors often use an elongated cylindrical structure to increase capacitor capacity. It extends along the height to form a three-dimensional volume. While within a limited mounting area, three-dimensional capacitors have much more surface area than planar capacitors, thus increasing the capacitance of the capacitor. However, the manufacturing cost increases further. Note that the area of a DRAM memory cell is typically 8–12F^2 as the minimum line width for

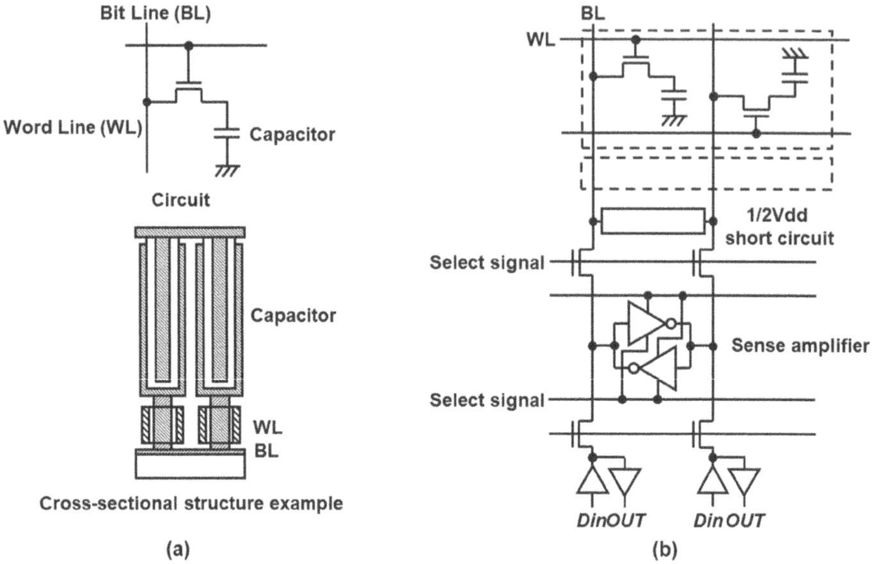

FIGURE 9.4
DRAM memory cells and basic array configuration. (a) DRAM memory cell and (b) DRAM data path.

microfabrication techniques using F. The small size is a major advantage. Compared to other memories, this will lower the cost per capacity.

Figure 9.4b shows the data path of a DRAM, which consists of row and column readouts. Memory cells are arranged in a two-dimensional array structure, and word lines run in the row direction. In row readout, the transistors in the cell array connected to the word lines selected by the address decoder are turned on, and data are read from all memory cells in that row simultaneously using the corresponding bit lines and sense amplifiers prepared for each bit line. The results are retained in the sense amplifier. In the subsequent column readout, this sense amplifier is accessed, and these data are read out in sequence.

The row read operation of DRAMs is also important for in-memory computing in Chapter 10 and is shown in Figure 9.5. Figure 9.5a shows the main circuit here. There is a cell capacitance *Cs* in the memory cell, and the bit line capacitance is *Cd*. The memory cell capacitance *Cs* is either *Vdd* or the ground potential *Vss* (0V), depending on the information being stored. Let this potential be *Vc* (*Vdd* or *Vss*). Before reading, the bit lines are precharged to 1/2 *Vdd* by the 1/2 *Vdd* short circuit in this figure and become floating. Also, the transistors in the memory cells are off. Then, as shown in Figure 9.5b, when the word line rises and the transistor in the specific memory cell connected to it turns on, the bit line is connected to the capacitor in the cell, and

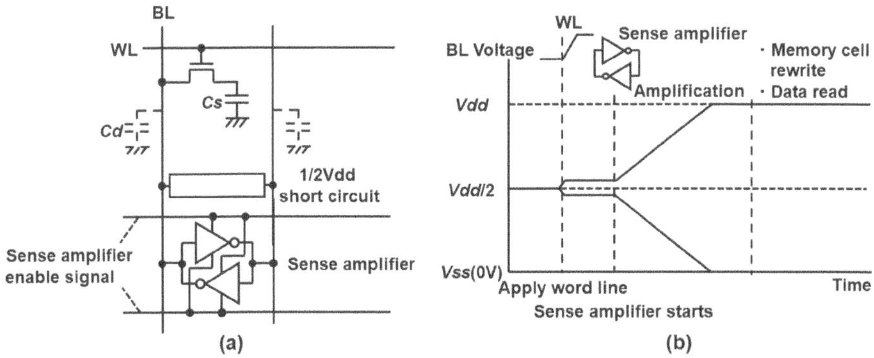

FIGURE 9.5
Basic DRAM read operation. (a) DRAM main circuitry and (b) DRAM read operation.

charge sharing is performed between the two. Therefore, the potential of the bit line becomes the value of the following simple formula, and a small signal appears on the bit line relative to 1/2 *Vdd*.

$$\text{Bit line Voltage} = \frac{Cd \times \frac{1}{2} Vdd + Cs \times Vc}{Cd + Cs}$$

Of the two bit lines input to the sense amplifier, one is at the above voltage and the other remains at 1/2 *Vdd*. The sense amplifier is then activated by the sense amplifier start signal, and a minute signal is amplified by the sense amplifier. This signal is read out as the memory information of the memory cell, as shown back in Figure 9.4b. At this time, the capacitor potential of the memory cell returns to its original value. This is called rewriting. In this way, each time a DRAM is read out, the information in the memory cell is destroyed and immediately written back in. However, since a large number of memory cells are connected to the bit lines, *Cd* is larger than *Cs*, and it is important to ensure the reliability of obtaining correct data by amplification during read. In this way, in DRAM, data are read from a specific memory cell using bit lines and word lines, and the signal is amplified by a sense amplifier and rewritten to read data and maintain the stored information.

Other DRAM technologies that use logic CMOS process technology are also being considered for DRAM by adding transistors. Figure 9.6a and b shows examples of 3T and 4T cells. The 3T cell is nondestructive and does not require rewriting because the contents of the cell are not lost when data are read. 4T cells use a latch to hold information. On the other hand, the Twin cell technology shown in Figure 9.6c combines two memory cells with a one-transistor, one-capacitor (1T1C) structure to store complementary information. This improves data reliability, enables low-voltage operation, and enhances soft-error tolerance.

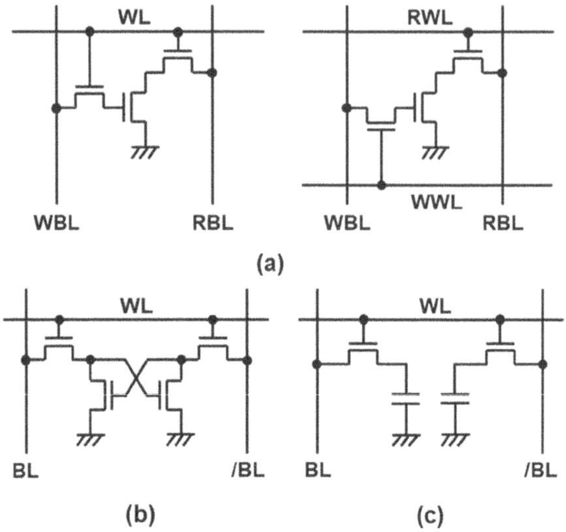

FIGURE 9.6
DRAM memory cell configurations. (a) 3T cells, (b) 4T cell, and (c) Twin T cell.

9.2 Nonvolatile Memory

9.2.1 Flash Memory

Flash memory [3,4] is a memory that stores information by creating a conductive area (floating gate) in the gate insulating film of a MOS transistor in which charge is stored and using the change in threshold voltage caused by the presence of charge in this area. Since the charge storage area is located in the insulating film, the charge in this area is retained even when electricity is cutoff, resulting in nonvolatile memory.

Readout is achieved by using a sense amplifier to detect that the channel current value of the MOS transistor varies with the threshold voltage. Charge storage is realized using the conductive regions surrounded by the insulating film or trap levels in the insulating film described above. To change the threshold voltage, charge is exchanged with the conductive floating gate region through the insulating film, through which no current normally flows. For this purpose, a high voltage, which is not used in normal memory, is used.

Figure 9.7a shows the circuit diagram and basic structure of a memory cell. The floating gate, which is characteristic of this memory, plays the role of storing electric charge. The presence or absence of charge (electrons) in the floating gate corresponds to 1s and 0s in the memory information. For example, the presence of a charge in the floating gate is set to "0", and the absence of a charge is set to "1. The control gate is an electrode for applying voltage. During writing, this electrode controls the flow of charge to and from the

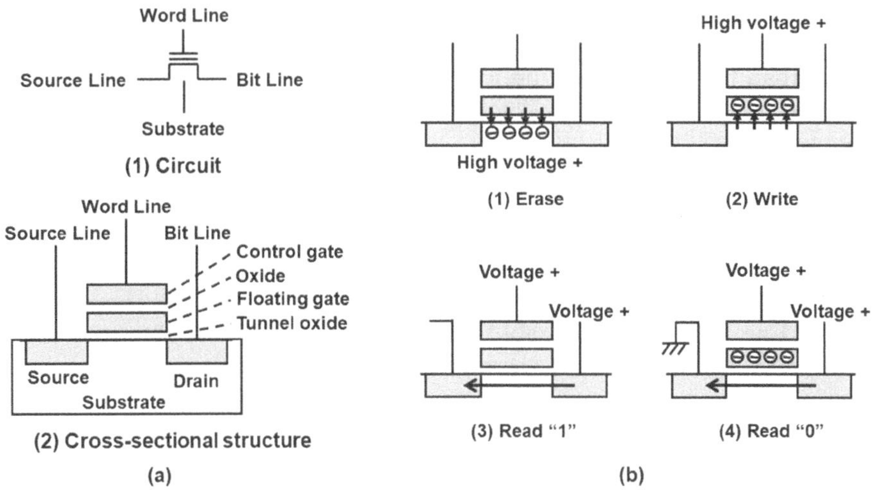

FIGURE 9.7
Flash memory cell and basic operation. (a) Memory cell and (b) basic operation.

floating gate and also performs readout and retention operations. Here, insulating films (oxide films) exist between the floating gate and the substrate and between the floating gate and the control gate. The oxide film between the floating gate and the substrate is specifically called the tunnel oxide film. This tunnel oxide film allows current to pass through as a tunnel current when a high voltage is applied. This current is used to transfer charge into and out of the floating gate. On the other hand, since the floating gate is sandwiched between insulating films, the leakage current is very small, resulting in a nonvolatile memory that retains its memory even when the power is turned off. Flash memory requires only a single transistor to form a memory element and to control it, making it small in area ($-4F^2$). It enables nonvolatile, high-density data storage.

The basic operation of a flash memory cell consists of erase, write, and read, as shown in Figure 9.7b. Erase (1) is performed to reset the information in the cell. In this process, a positive voltage is applied to the substrate, and a negative voltage is applied to the control gate. This process draws the charge in the floating gate to the substrate. Another method exists to draw electrons to the source. Historically, this operation was performed simultaneously for a large number of memories, hence the name flash memory. In (2), write, a negative voltage is applied to the substrate and a positive voltage is applied to the control gate. This operation injects electrons from the substrate into the floating gate. There is also a method that uses the phenomenon where a positive voltage is applied to the control gate and drain, and when electrons flow between the source and drain, some of them jump into the floating gate. In (3) and (4) read (Read), a positive voltage is applied to the control gate to determine the magnitude of the current between the source and drain.

This voltage is smaller than the voltage used for erase and write. The current flowing when no electrons are accumulated in the floating gate in (3) is larger than the current flowing when electrons are accumulated in (4). The difference in this current is detected by the amplifier to determine whether the memory cell is 0 or 1.

There are several flash memory schemes that differ in the rewriting method and connection method. Here, the NOR type and the NAND type are shown as the read data paths of the flash memory shown in Figure 9.8a. Both types use a sense amplifier to detect whether the current flowing in the selected memory cell is "0" or "1". In the NOR type, each memory cell is connected to a bit line. This allows for rapid bit-by-bit reading. This facilitates random access and is mainly used as a storage device for control programs. However, since write operations are also in the same unit, they are slow and not suitable for large capacities. Therefore, the NOR type is suitable for program storage in cell phones and routers. On the other hand, the NAND type has multiple memory cells connected in series, which is not suitable for bit-by-bit reading and is slow, so it is usually made to perform multiple bit lines as a unit. The same is true for writing, but since the write current is lower than that of the NOR type, increasing the degree of parallelism allows writing in units of a single data unit to be performed at high speed. Because it is suitable for high integration by connecting multiple memory cells in series, it is widely used as a storage device.

In flash memory, as shown in Figure 9.8b, a single memory cell can store many bits. The information was detected by changing the current that flows

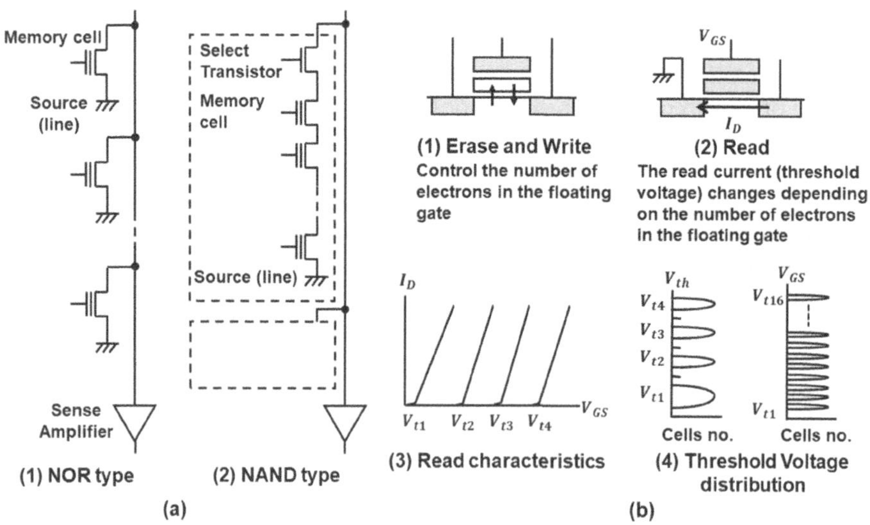

FIGURE 9.8
Read data path and multi-level technology. (a) Flash memory read data path and (b) multivalue technology.

during readout depending on the electrons in the floating gate, and this current changes with the number of electrons injected into the floating gate. Therefore, 2 bits of information can be stored in a single memory cell by controlling the number of electrons in the floating gate to create four current states, as shown in (3) of Figure 9.8b. Depending on the number of electrons, there are four threshold values as shown in (4) of Figure 9.8b. Currently, it is said that up to 16 values, or 4 bits of information, can be stored in a single memory cell, and commercialization of 2-bit and 3-bit memory cells has been achieved. In addition to the above, 3D stacking is also being used to increase capacity, with more than 200 layers now in mass production.

Flash memory is written by moving electrons in and out of a floating gate in an insulating film through a tunnel current by means of a high voltage. Therefore, the information in the memory cell can be read out at low read voltages, and this information is not lost when the power is turned off. However, there is a risk that the dielectric film may be destroyed by this operation.

This has been studied as a reliability issue for flash memory, and it is known that the tunnel oxide film can be destroyed by repeated writes and erases. Furthermore, there are phenomena called SILC, in which electrons in the floating gate escape through defects in the tunnel oxide film, as shown in Figure 9.9a, and Detrap, in which electrons enter the tunnel oxide film. SILC degrades the data retention characteristics of memory cells and is particularly problematic for long-term data retention. This phenomenon also widens the distribution of the threshold voltage of memory cells. Detrap becomes more pronounced with repeated write and erase cycles. The threshold voltage of

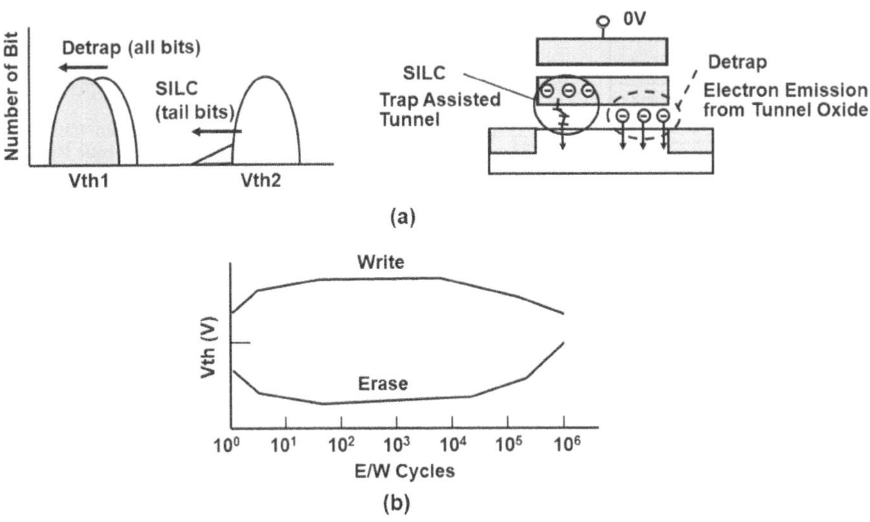

FIGURE 9.9
Memory cell reliability. (a) Threshold voltage fluctuation and (b) endurance memory window reduction.

the memory cell changes. In addition, flash memory has a limited number of rewrite cycles. As shown in Figure 9.9b, the threshold voltage corresponding to the two states of "1" and "0" changes as rewriting progresses, even when storing binary values.

These properties must be considered when using flash memory in artificial intelligence LSIs.

9.2.2 MRAM

Magnetoresistive random access memory (MRAM) is a thin magnetic film that uses spin orientation to store nonvolatile information. The number of rewrites is estimated to reach 10^{10} times or more. For learning functions, data must be rewritten frequently at high speed, and for this purpose, magnetic memory, with its high write endurance and low power consumption, is a promising nonvolatile working memory. Because it is a nonvolatile memory, it can also be applied to applications that eliminate leakage current in logic circuits during operation.

9.2.2.1 Current Induced Field Type MRAM and STT-MRAM

Among MRAMs, those that have reached the point of commercialization now (2025) are current-induced magnetic field MRAM [5] and spin–transfer–torque (STT) MRAM [6], as shown in Figure 9.10. Both of them use a sandwich structure called TMR as a memory element, in which a very thin insulating film is sandwiched between two ferromagnetic films, as shown in

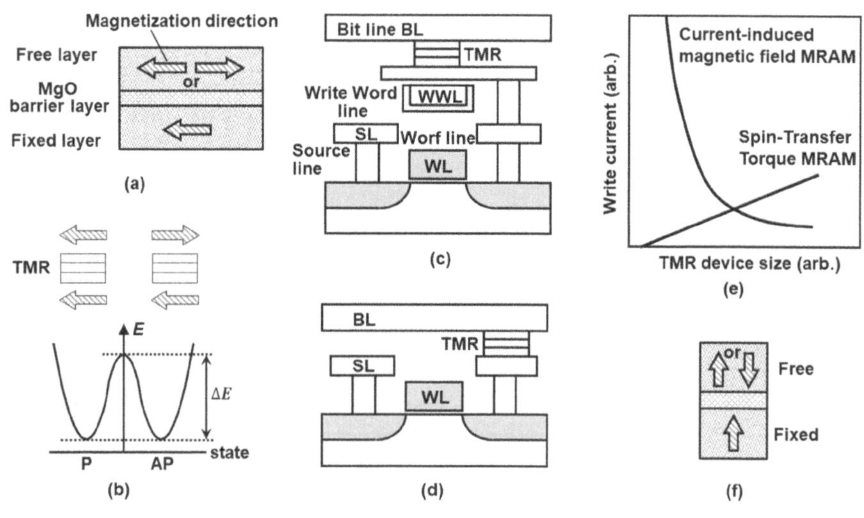

FIGURE 9.10
Magnetoresistive RAM (MRAM). (a) TMR device, (b) state of TMR device, (c) current-induced magnetic field MRAM, (d) spin-transfer torque MRAM, (e) dependence of write current on device size, and (f) perpendicular magnetization.

Figure 9.10a, which has a low operating voltage comparable to that of CMOS logic circuits and enables high-speed writing and a sufficient number of write cycles for learning. One of the two ferromagnetic films is called the fixed layer, and the direction of magnetization is fixed. The other is called the free layer, and several methods have been proposed to change the direction of magnetization.

Based on the relationship between the orientation of the magnetization of the fixed and free layers, the TMR device has the following two stable states, as shown in Figure 9.10b. In the parallel state (P state), the magnetization of both ferromagnets is oriented in the same direction. In this state, the electrical resistance of electrons passing through this device is low. In the antiparallel state (AP state), the magnetization of both ferromagnets faces the opposite direction. In this state, electrical resistance is high. In addition, there is an energy barrier ΔE between these two states, and this ΔE must be overcome to reverse the magnetization of the free layer, which is the rewriting of information. The size of this ΔE greatly affects the characteristics of the TMR device and is related to the rewrite current and thermal stability.

The current-induced magnetic field MRAM shown in Figure 9.10c began mass production in 2006. It has been put into practical use for industrial equipment and other applications. In this method, current is applied to the bit lines and write word lines, and the data are rewritten by the magnetic field generated around them. This method has the problem that it is difficult to increase capacity through miniaturization. In other words, as shown in Figure 9.10e, the finer the device, the larger the current required for rewriting becomes, and the inability to reduce the size of the transistor that supplies the rewriting current has been a bottleneck in miniaturization. In the STT-MRAM shown in Figure 9.10d, which was subsequently developed, data are rewritten using the current flowing to the memory element. The operating principle is to reverse the magnetization direction of the free layer of the TMR element using the torque effect of electron spin. The superiority of this method is that the current required for rewriting decreases with miniaturization, as shown in Figure 9.10e, making it easy to increase capacity through miniaturization. 2Mb experimental chips were introduced at an academic conference in 2007 [7], which stimulated development activity. This has led to miniaturisation and larger capacities, making it possible to replace some of the main memory, DRAM, and SRAM used for cache with nonvolatile memory. In particular, in recent years, major semiconductor manufacturers have been moving forward with mass production across the board.

Note that we have described two states of two ferromagnetic films: the parallel state (P state) and the antiparallel state (AP state). However, there is a choice between whether this orientation is parallel to the surface of the insulating film between the two films (Figure 9.10a) or perpendicular to it (Figure 9.10f). The one along the plane is easier to produce magnetization, and this one has been commercialized, but the perpendicular film has advantages such as improved thermal stability, so its development is progressing.

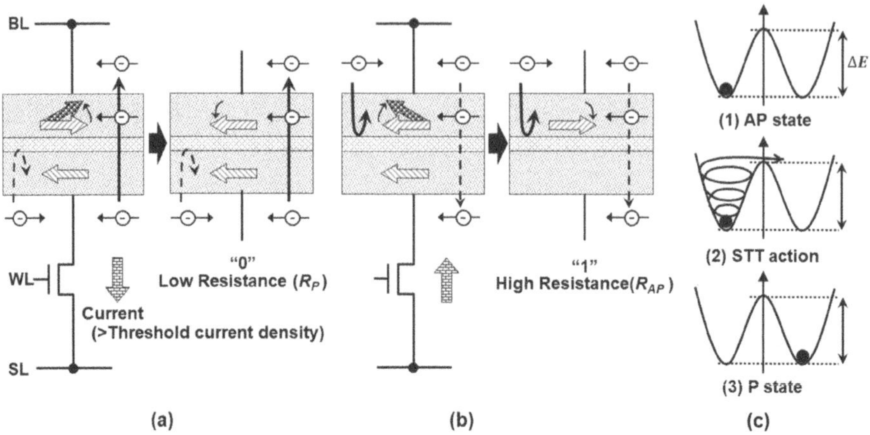

FIGURE 9.11
STT-MRAM write operation. (a) Write "0": parallelization, (b) write "1": antiparallelization, and (c) state change schematic.

Let us understand magnetic memory by looking at the write operation of STT-RAM, as shown in Figure 9.11.

In Figure 9.11, the memory cell consists of a tunnel magnetism (TMR) memory element and a selection transistor controlled by the word line WL The TMR memory element consists of a barrier a layer of MgO, an insulator, between two ferromagnetic layers, a fixed layer with a fixed magnetization and a free layer whose orientation can rotate freely. Considering the current path through the selection transistor from the bit line BL to the source line SL, if the direction of magnetization of these two ferromagnetic layers is parallel, the resistance in this area is low, and if it is antiparallel, it is high. In STT-MRAM, the direction of the current can create parallel or antiparallel magnetization in the two ferromagnetic layers. In other words, the information to be written is determined.

In the case of the parallelization (corresponding to information "0") written in Figure 9.11a, the current flows from the bit line to the source line, so the electrons flow in the opposite direction. Of these electrons, those with spin in the same direction as that of the fixed layer can pass through the barrier layer, but those with spin in the opposite direction are reflected. Therefore, electrons with spin orientations parallel to that of the fixed layer flow into the free layer. This affects (torques) the magnetization of the free layer, and when this current exceeds a certain threshold, the magnetization of the free layer becomes the same direction as that of the fixed layer, even if initially in the opposite direction. In this state, the resistance is as low as R_p.

On the other hand, in the case of antiparallelization (corresponding to "1" information) writing shown in Figure 9.11b, the current flows in the opposite direction from the source line to the bit line and electrons flow in the opposite direction. In the free layer, electrons with spin in the same direction as the fixed layer go directly to the fixed layer, but electrons with spin in the opposite direction to the fixed layer are reflected and exert a torque on the magnetization of the free layer. When this current exceeds the threshold current, the free layer becomes antiparallel to the fixed layer. The resistance R_{ap} at this point is higher than that of the parallel state R_p. This ratio in the following equation is called the TMR ratio:

$$\text{TMR ratio} = \frac{R_{ap} - R_p}{R_p}$$

The larger this is, the easier it is to read. The larger this value is, the easier it is to read out.

It should be noted that there is an energy barrier ΔE between P and AP, and it is necessary to overcome this ΔE in order to invert the magnetization of the free layer, which is the rewriting of information. In other words, as schematically depicted in Figure 9.11c, when a spin-polarized current passes through the free layer, the spin angular momentum of the electrons transfers to the magnetization of the free layer, generating a torque. This torque provides a force that rotates the magnetization of the free layer, causing a wobbling motion. When this wobbling motion crosses the energy barrier, the magnetization of the free layer reverses.

9.2.2.2 SOT-MRAM and VCMA-MRAM

MRAM is a nonvolatile memory with high write endurance and is a candidate for integration of working memory and file memory. Further improvements are underway to achieve this. Among these, MRAM that uses spin–orbit interaction to further reduce the write current is called spin–orbit torque MRAM [8], as shown in Figure 9.12a. This memory can mitigate the read disturbance in STT MRAM, where the write and read paths are the same. The memory element of SOT MRAM consists of a heavy metal layer, a magnetic layer, an insulator layer, and a magnetic layer, as shown in (1) of Figure 9.12a, i.e., a TMR element plus four layers of heavy metal layers. The difference between SOT MRAM and STT MRAM lies in the method of writing information to the cell. In SOT MRAM, writing is performed by passing a current horizontally through a heavy metal layer. The current flowing through the heavy metal layer causes a phenomenon called the spin Hall effect, which generates a spin current in the direction perpendicular to the current. This spin current injected into the free layer inverts the magnetization of the free layer. As shown in (2) of Figure 9.12a, the spin direction of this spin current can be changed by the direction of the current flowing in the heavy metal layer. This allows the P state and the AP state to be created in the TMR layer

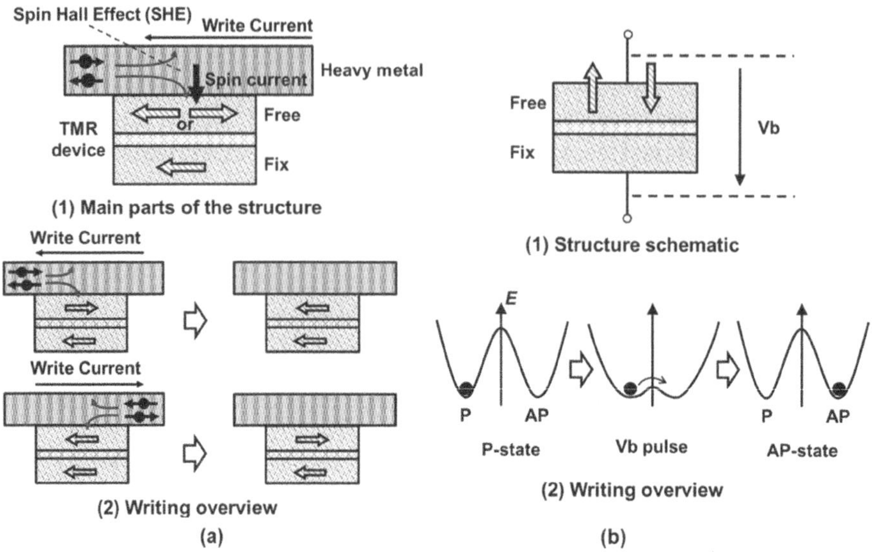

FIGURE 9.12
SOT-MRAM and VCMA-MRAM. (a) SOT-MRAM and (b) VCMA-MRAM.

according to the information to be written. During readout, the current flows in the direction of both ends of the heavy metal layer to prevent disturbances from occurring [9,10].

Furthermore, VCMA MRAM [11,12], as shown in Figure 9.12b, has been developed to reduce current and disturbance. In VCMA MRAM, voltage-controlled magnetic anisotropy (VCMA) is used to control the height of the energy barrier between the two states, as shown in (2) of Figure 9.12b, by adjusting the write voltage Vb during the write operation. When a voltage is applied, magnetization begins precession, but by controlling the pulse width, magnetization reversal can be achieved.

9.2.3 Ferroelectric Memory

Next is the ferroelectric memory (FeRAM) [13,14], which was also widely used in early transportation IC cards. Figure 9.13a shows the cross-sectional structure and circuit diagram, which consists of a selective transistor and a ferroelectric capacitor. In ferroelectric memory, the direction of residual polarization in this ferroelectric capacitor corresponds to the data value. Figure 9.13b shows the hysteresis of the voltage and polarization applied to the dielectric film in this capacitor. The polarity of the residual polarization can be changed by the polarity of the external voltage. This corresponds to the logic values "1" and "0".

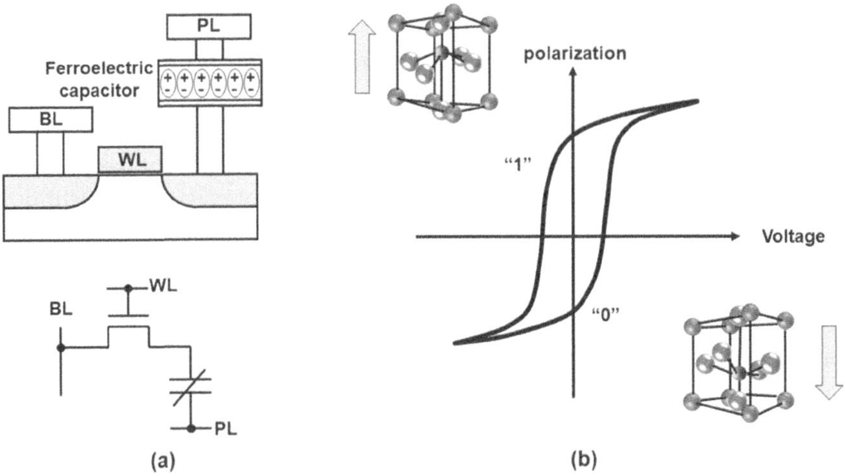

FIGURE 9.13
Ferroelectric memory. (a) Cross-section and circuit of memory cell and (b) polarization hysteresis curve.

When the application of voltage is stopped, residual polarization remains in the capacitor. This corresponds to a change in the position of the atoms in the ferroelectric body. The difference in this capacitance is read out. Each time a ferroelectric memory is read out. The information is destroyed and must be rewritten.

Unlike this figure, there is also a method in which a ferroelectric capacitor is placed under the gate of the MOS transistor, and the change in threshold value due to this is used as information.

The reliability of FeRAM is determined by the polarization characteristics of the ferroelectric capacitor. This characteristic changes over time. This degradation phenomenon of ferroelectric capacitors is called fatigue. This means that as the polarization reversal is repeated, the difference in hysteresis curves becomes smaller. In other words, the magnitude of the residual polarization (amount of charge) decreases. As the residual polarization decreases, both the signal that can be written and the signal that can be read decrease, which can result in write and read errors. Another phenomenon is imprinting. When imprinting occurs, the hysteresis curve shifts parallel to the voltage direction. For example, if the hysteresis curve shifts in the positive voltage direction, the residual polarization of positive charges decreases. This means that data cannot be stored there, which is a limiting factor for the data retention period. Imprinting is said to occur when the written data (direction of polarization) remains unchanged. As a result, the data retention characteristics are degraded.

9.2.4 Phase Change Memory

Phase-change memory [15] is a nonvolatile memory that uses the same material as an optical disk to store information by changing the crystal structure by different means, as shown in Figure 9.14a. In the amorphous state, the resistance is high, and in the crystalline state, the resistance is low. These two states as one bit are achieved by changing the current flowing through the memory cell to Joule heat at the electrodes and controlling the heating temperature and time variation of superheat caused by this change.

As shown in Figure 9.14b, to change to the high-resistance amorphous state corresponding to data '1', a large current is applied rapidly, cutoff after a short time, and the phase-change film is cooled rapidly. This creates an amorphous state, and data '1' are recorded. On the other hand, to change to the low-resistance crystalline state corresponding to data '0', the film is held at the crystallization temperature. A little crystalline state is formed by applying the required current over a long period of time, and data '0' is recorded.

In this memory, when the electrode area in contact with the phase-change material is small, the phase-change resistance can be changed with less power. This facilitates miniaturization. In addition, the phase-change resistance corresponding to "1" and "0" changes more significantly than that of MRAM, enabling high-speed read operations.

Furthermore, phase-change memory is becoming larger in capacity with the introduction of three-dimensional stacking technology. Crosspoint memory technology, which combines memory elements and selectors such as diodes, is attracting attention, and this is being stacked. Apart from this, multilevel technology is also being developed.

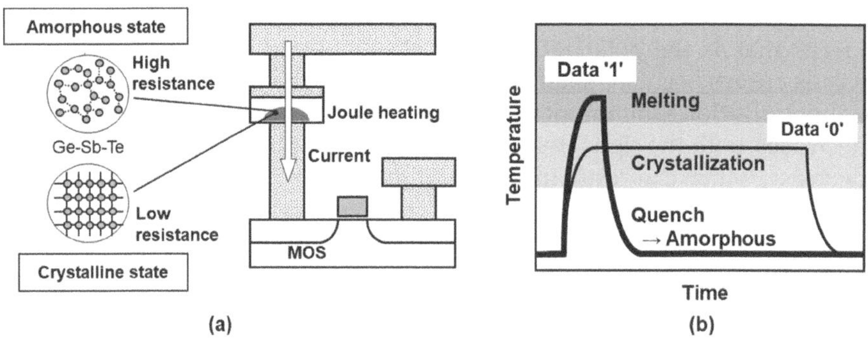

FIGURE 9.14
Phase change memory. (a) Structure of memory cell and (b) write method.

9.2.5 RRAM

Resistive random access memory (RRAM) [16] is a memory that uses the resistance change caused by the application of voltage (current). As shown in Figure 9.15a, the memory element of RRAM memory has a structure in which a thick insulator (oxide layer) is sandwiched between two electrodes. The characteristic feature of the memory element here is that its composition can be varied. As a general example, this insulator is a solid solution in which metal atoms are dissolved. The change in composition of this insulator changes its electrical resistance, enabling data to be stored and read.

Various models have been proposed for the principle of operation of RRAM. One of them will be the one shown in Figure 9.15a. The first is the initial state. No current paths are formed in the oxide layer. Once the memory device is fabricated, an additional operation called "forming" is usually required. This is the process of forming new filaments within the oxide layer by applying a predetermined high voltage because the initial resistance within the oxide layer is very high. This operation is performed only once for each memory element.

The operation of switching between the high-resistance and low-resistance states is controlled by the magnitude and direction of the applied bias. In the set operation, a specified voltage or current is applied to form a conductive path called a filament in the oxide layer. In the reset operation, the opposite voltage or current is applied to break the conducting path in the oxide layer. This increases the electrical resistance and data '0' is recorded. The filament is reset.

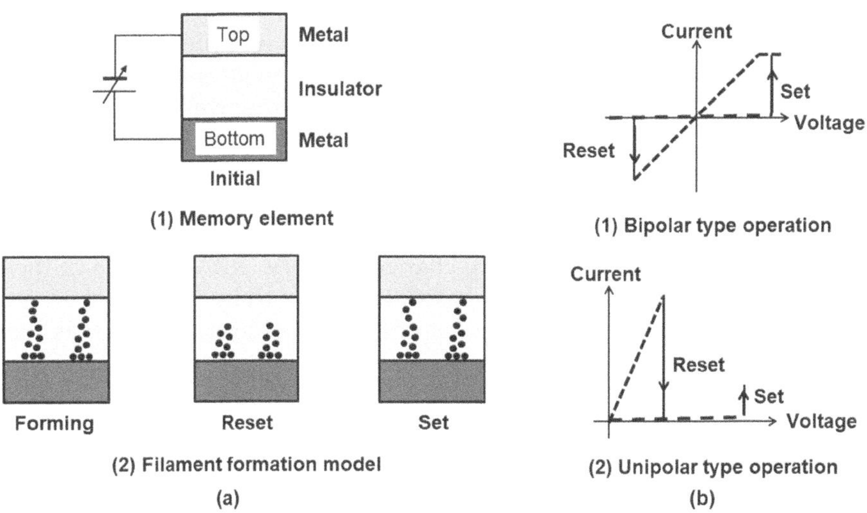

FIGURE 9.15
RRAM. (a) RRAM memory mechanism and (b) write method.

The RRAM operation can be divided into bipolar and unipolar operations based on the polarity of the reaction, as shown in Figure 9.15b. Bipolar operation depends on the polarity of the current and allows data to be written and read. Unipolar operation, on the other hand, is independent of the polarity of the current and can write and read data in only one direction.

RRAM is very simple. It is also capable of high-speed data write and read. Therefore, it is expected to replace NAND flash memory, although there are still issues with the number of rewrites.

In addition, there are ongoing attempts to enhance performance and streamline the hierarchical structure. Moreover, the development of a universal memory that integrates multiple memories is anticipated. Universal memory is defined as a memory that exhibits characteristics of both dynamic random-access memory (DRAM) and flash memory. It is anticipated that the device will exhibit DRAM-like high-speed write and read performance and will be nonvolatile and low in area, analogous to flash memory. This approach eliminates the necessity for a hierarchical memory structure, thereby enhancing system performance, reducing power consumption, and decreasing the number of components, consequently leading to smaller device dimensions.

9.3 Summary

This chapter describes the characteristics of major semiconductor memory technologies.

Memory is broadly classified into two categories: volatile and nonvolatile. Volatile memory encompasses two primary components: static RAM (SRAM) and dynamic RAM (DRAM). These components are primarily utilized for the temporary storage of data. Conversely, nonvolatile memory possesses the capacity to preserve information even in the absence of an active power source. In addition to flash memory, other memory technologies include magnetic RAM (MRAM), ferroelectric memory (FeRAM), phase-change memory (PCRAM), and resistance-change memory (RRAM). Each of these memory devices operates on the basis of a distinct physical principle, and their applications are categorized according to characteristics such as high speed, durability, and integration.

References

1. K. Osada, "Reviews and Prospects of Nanoscale SRAMs," *2006 IEEE International Conference on IC Design and Technology*, Padua, Italy, 2006, pp. 1–8. https://doi.org/10.1109/ICICDT.2006.220835
2. K. Itoh, *VLSI Memory Chip Design*, Springer, 2001.

3. F. Masuoka, M. Momodomi, Y. Iwata and R. Shirota, "New Ultra High Density EPROM and Flash EEPROM with NAND Structure Cell," *1987 International Electron Devices Meeting*, Washington, DC, USA, 1987, pp. 552–555. https://doi.org/10.1109/IEDM.1987.191485
4. C.-W. Yoon, "The Fundamentals of NAND Flash Memory: Technology for Tomorrow's Fourth Industrial Revolution," *IEEE Solid-State Circuits Magazine*, vol. 14, no. 2, pp. 56–65, 2022. https://doi.org/10.1109/MSSC.2022.3166466
5. M. Durlam, B. Craigo, M. DeHerrera, B. N. Engel, G. Grynkewich, B. Huang, J. Janesky, M. Martin, B. Martino, J. Salter and J. M. Slaughter, "Toggle MRAM: A Highly-Reliable Non-Volatile Memory," *2007 International Symposium on VLSI Technology, Systems and Applications (VLSI-TSA)*, Hsinchu, Taiwan, 2007, pp. 1–2. https://doi.org/10.1109/VTSA.2007.378942
6. M. Hosomi, H. Yamagishi, T. Yamamoto, K. Bessho, Y. Higo, K. Yamane, H. Yamada, M. Shoji, H. Hachino, C. Fukumoto and H. Nagao, "A Novel Nonvolatile Memory with Spin Torque Transfer Magnetization Switching: Spin-Ram," *IEEE International Electron Devices Meeting, 2005. IEDM Technical Digest*, Washington, DC, USA, 2005, pp. 459–462. https://doi.org/10.1109/IEDM.2005.1609379
7. T. Kawahara, R. Takemura, K. Miura, J. Hayakawa, S. Ikeda, Y. Lee, R. Sasaki, Y. Goto, K. Ito, T. Meguro and F. Matsukura, "2Mb Spin-Transfer Torque RAM (SPRAM) with Bit-by-Bit Bidirectional Current Write and Parallelizing-Direction Current Read," *2007 IEEE International Solid-State Circuits Conference. Digest of Technical Papers*, San Francisco, CA, USA, 2007, pp. 480–617. https://doi.org/10.1109/ISSCC.2007.373503
8. K. Jabeur, L. D. Buda-Prejbeanu, G. Prenat and G. di Pendina, "Study of Two Writing Schemes for a Magnetic Tunnel Junction Based On Spin Orbit Torque," *World Academy of Science, Engineering and Technology, International Journal of Electrical, Computer, Energetic, Electronic and Communication Engineering*, vol. 7 pp. 1054–1059, 2013. https://doi.org/10.14810/elelij.2017.6101
9. H. Kazama and T. Kawahara, "Spin-Orbit Torque MRAM Read Reliability," *2017 IEEE International Magnetics Conference (INTERMAG)*, Dublin, Ireland 2017, pp. 1–2. https://doi.org/10.1109/INTMAG.2017.8007571
10. Y. Kishi, A. Yamada, M. Ke and T. Kawahara, "Examination of Magnetization Switching Behavior by Bi-Directional Read of Spin-Orbit-Torque MRAM," *IEEE Transactions on Magnetics*, vol. 58, no. 5, pp. 1–10, 2022, Art no. 3400510. https://doi.org/10.1109/TMAG.2022.3154025
11. W. Kang, Y. Ran, W. Lv, Y. Zhang and W. Zhao, "High-Speed, Low-Power, Magnetic Non-Volatile Flip-Flop With Voltage-Controlled, Magnetic Anisotropy Assistance," *IEEE Magnetics Letters*, vol. 7, pp. 1–5, 2016, Art no. 3106205. https://doi.org/10.1109/LMAG.2016.2604205
12. T. Nozaki, T. Yamamoto, S. Miwa, M. Tsujikawa, M. Shirai, S. Yuasa, and Y. Suzuki. "Recent Progress in the Voltage-Controlled Magnetic Anisotropy Effect and the Challenges Faced in Developing Voltage-Torque MRAM," *Micromachines*, vol. 10, no. 5, p. 327, 2019. https://doi.org/10.3390/mi10050327
13. S. Kawashima, K. Morita, M. Nakazawa, K. Yamane, M. Ogai, K. Kawabata, K. Takai, Y. Fujii, R. Yasuda, W. Wang, Y. Hikosaka, and K.' Inoue, "An 8-Mbit 0.18-μm CMOS 1T1C FeRAM in Planar Technology," *IEICE Transactions on Electronics*, vol. E98-C, no. 11, pp. 1047–1057, 2015. https://doi.org/10.1587/transele.E98.C.1047

14. T. Francois, J. Coignus, A. Makosiej, B. Giraud, C. Carabasse, J. Barbot, S. Martin, N. Castellani, T. Magis, H. Grampeix and S. Van Duijn, "16kbit HfO_2:Si-Based 1T-1C FeRAM Arrays Demonstrating High Performance Operation and Solder Reflow Compatibility," *2021 IEEE International Electron Devices Meeting (IEDM)*, San Francisco, CA, USA, 2021, pp. 33.1.1–33.1.4. https://doi.org/10.1109/IEDM19574.2021.9720640
15. H.-S. P. Wong, S. Raoux, S. Kim, J. Liang, J. P. Reifenberg, B. Rajendran, M. Asheghi and K. E. Goodson, "Phase Change Memory," *Proceedings of the IEEE*, vol. 98, no. 12, pp. 2201–2227, 2010. https://doi.org/10.1109/JPROC.2010.2070050
16. H.-S. P. Wong, H. Y. Lee, S. Yu, Y. S. Chen, Y. Wu, P. S. Chen, B. Lee, F. T. Chen and M. J. Tsai, "Metal-Oxide RRAM," *Proceedings of the IEEE*, vol. 100, no. 6, pp. 1951–1970, 2012. https://doi.org/10.1109/JPROC.2012.2190369

10

AI Computing with Semiconductor Memory (In-Memory Computing)

Normal information processing uses a separate architecture for computation and memory. The processor is in charge of information processing operations, while the memory functions as a storage location for information. For this reason, data are transferred between the storage device (memory) and the computing device (CPU) in the hardware execution of information processing operations. However, as mentioned in Chapter 9, data transfer consumes a lot of power and is slow. In addition, the operating speed of the arithmetic unit is high, while that of the memory unit is low.

For the purpose of improving the performance of information processing and reducing power consumption, the integration of storage devices and arithmetic devices is being promoted to solve this problem. In particular, in the operation of a neural network, which is an artificial intelligence process, many parts are connected to perform a sum-of-products operation using the weights stored in the memory device and the results of the previous stage operation. Therefore, it is suitable for reducing data transfer between the storage device and the arithmetic unit by incorporating the arithmetic mechanism into the memory. In addition, combining it with nonvolatile memory, which is under development, will further reduce power consumption.

10.1 Artificial Intelligence Processing and Memory

In information processing systems at the edge, it is desirable to reduce power consumption while maintaining high performance. On the other hand, in today's information processing systems, as shown in Figure 10.1a, there is a large performance gap [1,2] between the CPU (processor) and memory that performs operations. The processor is getting faster, but the memory remains slow. This is due to the fact that memory cell arrays take a two-dimensional arrangement and have a fixed area. With each successive generation, processing technology becomes finer and finer, but at the same time, higher capacity is required. Therefore, the electrical delay time of that signal cannot be reduced in a 2D memory array suitable for the data size used in that generation. As shown in Chapters 8 and 9, in memory, there are metal wires

DOI: 10.1201/9781003599067-10

running vertically and horizontally on the 2D memory array, and the memory cell selected at the intersection of these wires is read out. Row-by-row transfers also increase in their capacity. In other words, while CPU performance improves with miniaturization, memory performance is difficult to improve because of the increased capacity, and as a result, the performance ratios are diverging.

Since the memory and CPU are created as separate blocks or chips, data transfer between the memory and CPU also takes time, as shown in Figure 10.1b. When configured as separate chips, the round-trip signal delay is nearly an order of magnitude larger than the signal delay inside the chip. Furthermore, the power consumption associated with this signal transmission is also an order of magnitude larger, sometimes three orders of magnitude larger than the power for the signal processing operations themselves. The fact that access to memory is a bottleneck in system performance from the CPU's point of view has long been known as the von Neumann bottleneck.

In these issues, the hierarchical structure of memory shown in Chapter 8 was taken as one of the solutions to Figure 10.1a (the structure includes flash memory, which is even slower storage). This chapter deals with solving the issue of data transfer between the CPU and memory itself, as shown in Figure 10.1b. This will be a method suitable for artificial intelligence LSIs.

The concept itself is simple. The latency of data transfer between memory and CPU and the power consumption are simply solved by shortening the distance of data transfer between memory and CPU as much as possible. The first step in achieving this is near-memory computing. This is particularly effective in reducing power consumption. Artificial intelligence LSIs used at the edge are required to be used in environments with limited power supply capacity. Under a certain power supply performance, more power can be allocated to computation instead of data transfer, and as a result, system performance can be improved.

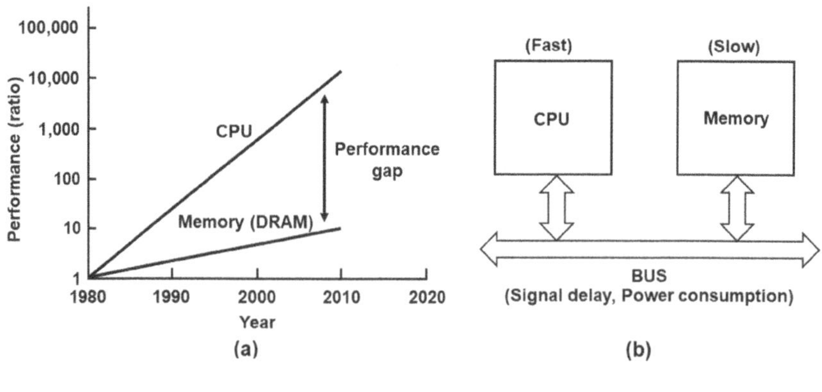

FIGURE 10.1
(a) CPU and memory performance and (b) data transfer between CPU and memory.

Furthermore, there are attempts to completely remove the interface between the CPU and memory. This is in-memory computing. Here, the computation is embedded in the core of the memory cell array. In memory readout, as we have seen in previous chapters, when a word line is selected in a memory cell array, memory cell information is read out using multiple bit lines. As shown in Figure 10.1b, this operation is slow compared to the CPU. Furthermore, it consumes a lot of power to drive a large number of bit lines. In conventional memory, this time and power are used only to read data from the memory, which is natural since it is a memory. In-memory computing incorporates arithmetic functions.

Note that in these near-memory and in-memory computing, the signal is processed near where the data physically resides. For this reason, it is also called data-centric processing.

10.2 Near-Memory Computing

10.2.1 Composition and Effects

The concept of near-memory computing [3] is illustrated in Figure 10.2. Figure 10.2a shows a conventional computing configuration with a hierarchical memory structure. To solve the issue in Figure 10.1 with this configuration, it was necessary to increase the capacity of the cache memory. Furthermore, although omitted in Figure 10.2a, this cache memory itself is also prepared as primary, secondary, and even tertiary in some large systems. At this time, as the number of processor cores increases, the secondary cache memory common to these cores and below must store the information that is expected and predicted to be used by the various cores. This also led to an increase in power and cost. In other words, data in cache memory must be transferred from lower-level memory or storage if it cannot be used for the operation at that time (in the event of a cache miss). This requires additional power and time.

Near-memory computing, shown in Figure 10.2b, on the other hand, aims to process data close to the physical location where it resides. It performs the necessary processing directly connected to main memory or even directly to storage. Thus, a large cache memory for the entire system is unnecessary (the concept of cache misses does not exist, so to speak), and efficient computation in terms of both power and speed can be achieved. And, as we have seen, the calculations performed by the artificial intelligence processing LSI are data-driven signal processing, and the data to be used are prepared in memory in advance, making this method of processing effective. The distance of data transfer between the memory and the processor can be shortened, and the performance improvement and low power consumption of the artificial intelligence processing LSI can be achieved.

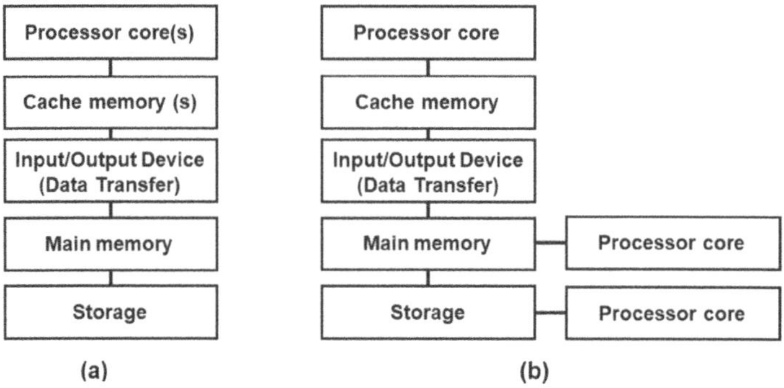

FIGURE 10.2
Basic architecture of near memory computing. (a) Conventional computing architecture and (b) near-memory computing architecture.

The main function required at the edge is inference, and the weight data, etc. used in the calculation of the neural network that constitutes the artificial intelligence LSI are fixed, and these can be read from the nonvolatile memory. In addition, even in the case of learning described below, updating the weights (parameters), i.e., memory rewriting, is infrequent and can be done within the limit of the number of rewrites of the nonvolatile memory. Furthermore, depending on the usage and the type of memory used, the data itself can also be written to nonvolatile memory. Although this is not a technology specific to memory itself, it is possible to configure a system that supplies power only during the performance of calculations and cuts off power at other times.

The use of this nonvolatile memory also reduces the size and cost of devices. As seen in the previous chapter, as volatile memory, SRAM is compatible with logic processes but has a large area (120–150F^2), and DRAM has a small area (8–12F^2) but the manufacturing process is complex and costly (F is the minimum line width that can be produced using the microfabrication technology used). Therefore, the use of nonvolatile memory, which is less process-complex than DRAM and smaller in area than SRAM, can be effective when relatively large capacities are required. This is also important for in-memory computing, which is discussed in the next section.

Let's look at some techniques for near-memory computing.

10.2.2 Large-Scale Embedded Memory

Although large-scale embedded memory devices are under active development, it is purposely mentioned here because artificial intelligence processing LSIs are data-driven signal processing. In other words, the solution is to embed as much large-capacity memory as possible. This can be called cache memory in a sense, but there are no miss-hits. All the contents necessary for the calculation are available. Thus, large-scale embedded memory can provide the processor with the necessary data instantaneously, greatly increasing the processing speed of the computer. It also reduces power consumption. Since calculations performed by artificial intelligence processing LSIs require processing large amounts of data at high speed, large-scale embedded memory is important.

However, there are several challenges in realizing large-scale embedded memory. First, memory cells and array structures must be as small and stable as possible. As memory capacity increases, the operation via the internal bus may also slow down the overall performance. Increased cost is also a concern. This is because not only does the chip area need to be increased to realize large-scale embedded memory but also additional steps are needed in the manufacturing process to match the structure of the memory cells. The difference between in-memory computing and the in-memory computing described in the next section is that the computation and memory sections are integrated into a single unit.

10.2.3 3D Implementation of Processor/Logic Circuits and Memory

10.2.3.1 Integration of Processor/logic Circuit Chips and Memory Chips in a Package or on a Board

As shown in Figure 10.3a, having chips mounted also in the height direction [4], either in a package or on a small board, improves performance in multiple ways, including speed and power, compared to a two-dimensional arrangement of processor/logic chips, memory chips, and interface chips on a regular board. Processor/logic chips play a central role in data processing, performing calculations and control. Computing speed and efficiency are required, while memory chips temporarily store data and quickly provide the information needed by the processor. The balance between speed and

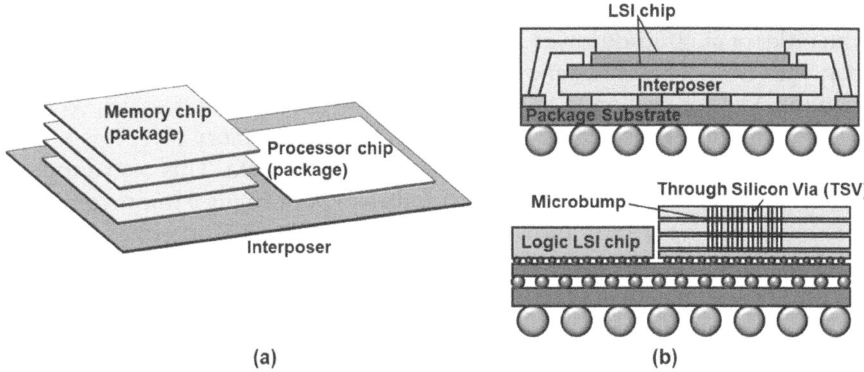

FIGURE 10.3
3D implementation of processing circuits and memory. (a) Stacking and integration on a board and (b) 3D stacking.

capacity was important. By placing these chips close together, data transfer delays can be minimized and power consumption reduced. This maximizes performance. To achieve this, interposers are used to compactly place multiple chips. Reducing the implementation area is important when placing chips on the edge. Achieving an optimal balance between size and performance is important. Here, an interposer is a substrate for efficient connection between different semiconductor chips. It also serves to effectively disperse the heat generated by multiple chips and help manage temperature. Research and development of new materials, such as organic interposers and glass interposers, are also underway.

What is particularly important in terms of performance improvement in three-dimensional (3D) mounting [4] is the ability to connect chips together using short metal wires (wires) by utilizing the height direction. In conventional 2D mounting (planar mounting), long wires must be used, resulting in delays associated with interconnections, but in 3D mounting, these delays can be significantly reduced. Shortening the wire length increases the signal transmission speed and speeds up the data transfer between the arithmetic circuit and the memory.

In addition, the height direction can be utilized to optimize each block of chips that are divided into multiple pieces. This has the advantage that arithmetic circuits and memory can be placed in different layer structures according to performance requirements. In other words, circuits optimized for specific tasks can be placed where they are needed, thereby optimizing the performance of the overall system.

10.2.3.2 Stacked Structure of Arithmetic Circuit Layer and Memory Layer

Figure 10.3b is more direct, with the processor/logic circuit layer and memory layer stacked as layers of silicon. In particular, in the case of DRAM, multiple memory layers and one logic circuit layer are stacked, and technologies such as through-silicon vias (TSVs) and microbumps are used to connect them. Power is an order of magnitude lower, and bandwidth is an order of magnitude higher than with DDR3, the conventional electrical specification for chip-to-chip connections. The connection to the processor/logic circuit layer is easier, so there is less delay in this aspect as well. In addition, the memory layer has a multibank structure that allows simultaneous access to data. This also supports highly parallel processing, which contributes to performance improvement.

This stacked approach allows for customization of the interface with the processor/logic circuit layer while still using regular DRAM cells, making the design suitable for specific applications.

These precautions require proper thermal management because heat tends to concentrate with high integration. Failure to do so can cause device performance degradation and failure. In addition, advanced interposer and 3D integration technologies can raise manufacturing costs. High-performance chips for edge devices must be cost-effective.

10.2.4 Examples of Standards

The development of three-dimensional stacking of these layers of processors and logic circuits with layers of memory chips to achieve high-speed data access and efficient data transfer has been active, and a number of standards exist, as described below.

10.2.4.1 Hybrid Memory Cube (HMC) [5]

It offers more than 20 times the bandwidth of conventional DDR memory, significantly improving data transfer rates and processing performance. It is also highly energy efficient. Individual DRAM chips are stacked vertically, and multiple chips are electrically connected using Through-Silicon Via. Although HMC is widely used in high-performance computers and data centers, heat concentration is still an issue and cooling techniques must be optimized.

10.2.4.2 Wide-IO [6]

It is a 3D stacked memory standard designed for mobile devices. It improves data transfer speeds by using multiple data lines simultaneously. Compared to conventional low-power double data rate (LPDDR) standard memory, it has a data transfer rate that is up to twice as fast. It combines low power consumption with high bandwidth to achieve both high-speed data access and power savings. In mobile devices, it plays an important role in extending battery life while running high-performance applications.

10.2.4.3 High Bandwidth Memory (HBM) [7]

It is the leading high-bandwidth memory standard. It is used in graphics cards and high-performance computing. Processors and memory are stacked in 3D and connected using TSVs. It has a bandwidth of up to 1 TB/s or more and can achieve high data transfer rates and efficient memory access.

10.3 In-Memory Computing

A technology that enables in-memory processing, reducing data movement and improving computational efficiency.

10.3.1 Basic Configuration and Operation

In-memory computing or computing-in-memory [8,9] has recently attracted attention as an energy-efficient parallel processing arithmetic computer architecture. In particular, it plays an important role in edge computing to realize artificial intelligence. The key to neural network processing is the sum-of-products operation, and the performance and power of this part determine the entire system. In in-memory computing, this product-sum operation is implemented by embedding the operation described in Chapter 5 for two-dimensional arrays into memory cell arrays, utilizing the fact that memory cells form arrays in two dimensions. Since the two-dimensional array of memory cells is a common memory configuration, this basic configuration is not dependent on the type of memory cell.

Let's use the example of the sum-of-products operation in RRAM described at the end of the previous chapter.

As shown in Figure 10.4a, consider the input/output X_i, Y_i of a neuron as binary (0 or 1), which, using the weights W_{ij}, results in the sum-of-products $u_i u_i$ is the output Y_i via the activation function f. In this figure, the number of inputs is n and the number of outputs is 4. Let us consider the processing of this single-layer neural net as in-memory computing. This processing can be embedded in a two-dimensional array of memory cells.

Here, we also take advantage of the fact that RRAM memory cells can store analog values according to the value of their internal resistance. In other words, the value of the resistance of the RRAM memory cell is used as the weight value. The memory cells are arranged in a two-dimensional array as shown in Figure 10.4b, with a binary input (0 or 1) of X_i applied to each row and used as the multiplier for all RRAM memory cells in the same row. The value of the weights is the value of the resistor elements within each RRAM memory cell. When a word line is selected, the input and the weights of the resistive elements are multiplied. The result of this multiplication appears on the bit lines for each column, where the outputs from the many memory cells in this column are added together. In other words, this means that a

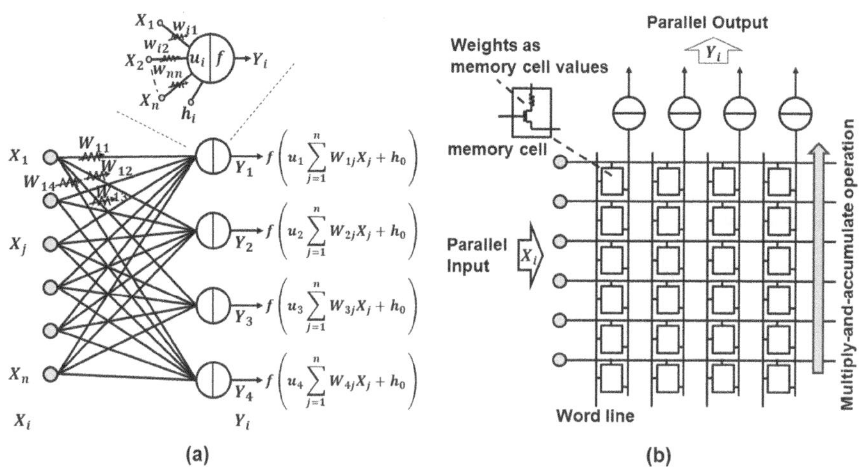

FIGURE 10.4
Basic architecture of in-memory computing. (a) Processing in a single layer of neural network and (b) computation using 2D memory array.

sum-of-products operation has been performed. Moreover, this is done in parallel with the number of bit lines. Ideally, all inputs are applied simultaneously to obtain the output. Thus, a massively parallel sum-of-products operation is performed.

Here is an example where the activation functions are also all provided in parallel. The result can be the output of the next layer of the neural network. The same configuration can be used there as well, using a two-dimensional array of memory cells. As explained in the columnar readout of memory, this can be done as a data transfer path using a single activation function, which is done in sequence in multiple rows.

Figure 10.5 will be used to illustrate Figure 10.4b separately. It shows how the sum-of-products operations of a neural network in one layer are mapped

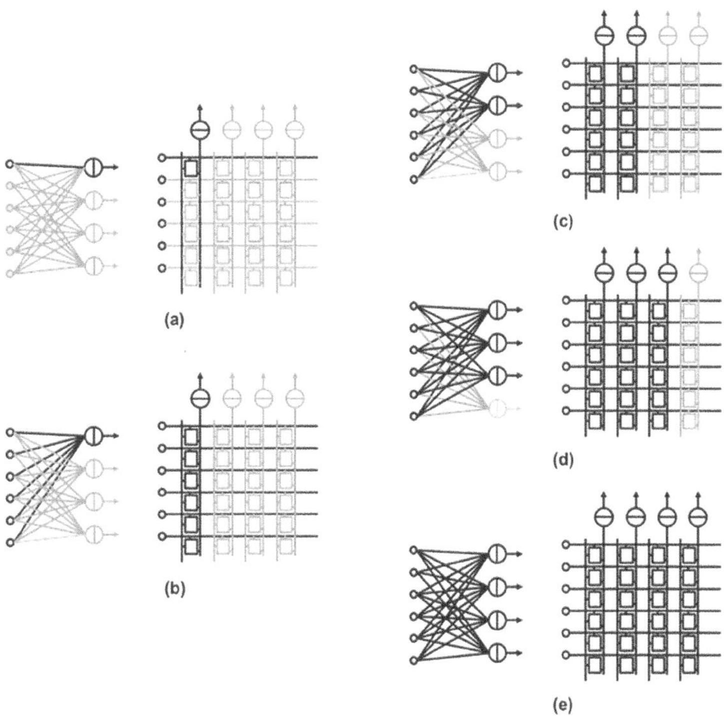

FIGURE 10.5
Neural networks embedding. (a) One of the inputs to the first neuron and its corresponding location in the memory array. The memory cell multiplies the input by the resistor's weight, and the result appears on the bit line. (b) All the inputs to the first neuron and their correspondence on the memory array. A product-sum calculation is performed using one column of memory for one neuron, and the result becomes the input to the activation function. (c) A correspondence on the memory array with the second neuron added. Using all of the same inputs, a calculation is performed for this one neuron using the corresponding memory column, and the result is input to the activation function. (d) A correspondence on the memory array with the third neuron added. (e) A correspondence on the memory array with all neurons added. The neurons in one layer have been mapped onto the memory array.

to a two-dimensional memory array of in-memory computing. Let us assume that this layer has six inputs and is fully coupled to four neurons.

When a word line is selected, as shown in (a), for one of the inputs, the product of the input and the weights of the resistors is taken, and the result appears on the bit lines. As shown in (b), this product is done on a single word line, which means that all the inputs considered here have been multiplied by their corresponding resistor weights, all of which appear on the bit lines. Here, the outputs from all the inputs are added together in this column. This single column represents the sum-of-products calculation for a single neuron, which is the input to the activation function. Thus, as shown in (c), the next column represents the next neuron, where the same sum of products is performed using all the inputs and input to the activation function. This is followed by (d) and (e), where in all combinations of six inputs and four neurons in a layer, the product of all inputs and resistor weights is taken for each neuron, and this result is summed for each neuron and input to the respective activation function. In other words, this completes the computation in this layer.

The arithmetic process in a neural network is based on having multiple inputs and weights, taking the product of the corresponding two, and summing the whole. Furthermore, the desired function is achieved by combining multiple identical circuits in this way. Thus, in-memory computing can be embedded in a two-dimensional memory array and operate in parallel. This makes it an energy-efficient process with a small footprint, which is especially important for edge computing, where power supply limitations exist.

Next, we will look at the characteristics of in-memory computing using each type of memory.

10.3.2 Configuration with SRAM

SRAM can be created in a logic process and does not require refresh like DRAM, making it easy to read and write operations. However, the memory cell area of SRAM is larger than that of DRAM, etc., so the level of integration is low. The information that can be stored is one bit (binary). In addition, the operation of SRAM has become unstable with the progress of miniaturization and lower voltages.

This in-memory computing using SRAM [10] as a sum-of-products operation is performed as follows.

Figure 10.6a shows the simplest configuration. Each memory cell has one bit of information, and the selection of the word line is based on the input one-bit information. In other words, when the input is "1", the word line has a positive pulse, and when the input is "0", the word line does not change. With this configuration, when an input (pulse or DC) is applied to the word line, a binary multiplication is immediately performed in the SRAM cell, and the result appears in the bit lines based on the input and stored weight values. By operating multiple word lines, a cumulative voltage drop of the number

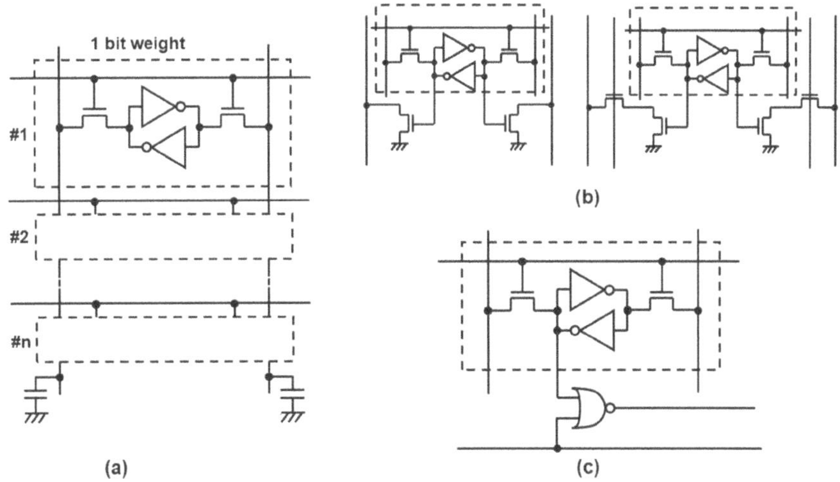

FIGURE 10.6
SRAM-based architecture. (a) Basic circuits structure using SRAM, (b) example of a multi-transistor SRAM cell, and (c) 1-bit weight digital circuit.

of memory cells will occur in the bit lines. In other words, each memory cell holds specific bits, and these bits are combined to represent multiple bit weights. In operating multiple word lines, the time to activate each word line may be varied. In such an operation, the bit lines sum or redistribute the charges, and if this is done in a linear fashion, it can be retrieved as an arithmetic result. However, the operating range must be within a voltage range and accuracy that allows detection of the potential difference in the bit lines. In addition, the threshold voltage of the nMOS transistors connected to the bit lines must be set so that it does not affect the bit line drops. This can be improved by using an SRAM cell with 8–10 transistors, as shown in Figure 10.6b, instead of the usual SRAM cell with six transistors. In this way, memory cell read operations and arithmetic operations can be performed simultaneously. And, as is true of in-memory computing as a whole, this is done simultaneously with multiple bit lines that connect to memory cells corresponding to word lines. Moreover, the number of such cells is large, such as 128 or 256. In other words, 128 parallel operations or 256 parallel operations. This is the reason why in-memory computing is expected to improve the performance of AI chips. In order to improve parallelism, memory cells of 6T or more are being used, which increases the area of memory cells per cell, but improves computing performance by increasing parallelism.

The summation operation can be performed digitally instead of analogously by summing or redistributing the charges at the capacitance. The unit of configuration in this case is shown in Figure 10.6c. This is a unit of configuration using a standard 6T SRAM cell and a NOR gate that functions as a bit-by-bit multiplier. Using this, as shown in Figure 10.7a, a configuration

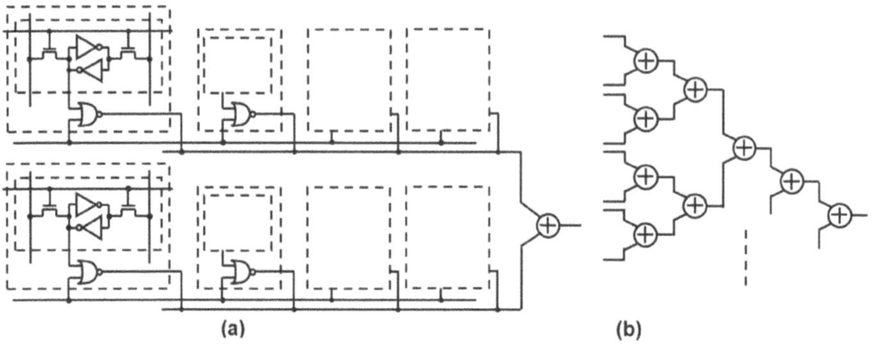

FIGURE 10.7
Multibit digital circuit configuration example. (a) Digital circuit configuration using SRAM weight cells and (b) large-scale configuration.

can be realized that takes the product of a signal whose weights are 4 bits as an example, with the input as binary, and the sum of the product of this, another input, and another 4-bit weight connected to it. Although the circuit scale becomes slightly larger based on this configuration, as shown in Figure 10.7b, it is possible to calculate the sum of products of these signals and weights for multiple inputs.

The challenge is that this circuitry is a large area and energy-consuming block when done digitally. Analog operations, such as charge redistribution using capacitance, are very efficient because they are simply processed on bit lines. On the other hand, the digital approach can easily benefit from miniaturization and is more stable against noise.

10.3.3 Configuration with DRAM

DRAM also stores one bit of information per memory cell. The cell area is small, and a large capacity is possible. This is arranged in two dimensions, and word drivers and sense amplifiers are arranged to form a memory array. Configuration using DRAM [11] attempts to perform operations on each of these bit lines. By incorporating neural network operations into these operations, it is an attempt to achieve high computing performance in a small footprint. Since the level of integration is high, the chip is divided into multiple chunks and handled as a bank that can be independently controlled within a single chip. There is a method in which a logic circuit is installed in each of these banks to perform operations as a neural network. In addition, there are many developments with a three-dimensional structure in which logic chips are stacked on top of each other. This is rather similar to near-memory computing, but the aim is to create a method that combines memory and logic.

The method reported for bit line-by-bit line operations is a method that uses a normal DRAM memory cell array configuration, as shown in Figure 10.8a, or a memory cell array composed of multiple transistors similar to early

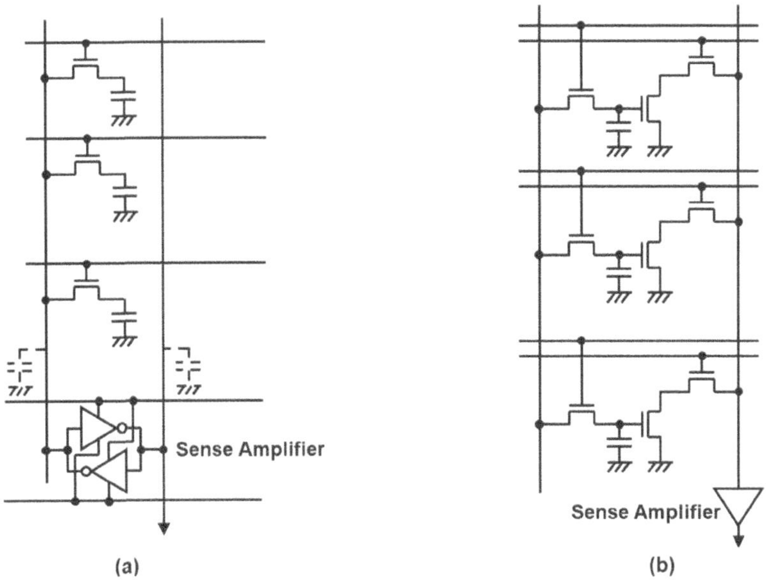

FIGURE 10.8
Bit line-based operations in DRAM memory cell arrays. (a) Computation using DRAM cells and (b) computing using gain-type DRAM cells.

DRAM memory cells, in which the write and read operations are separated, as shown in Figure 10.8b. As shown in 10.8 (b), this method performs operations by selecting multiple word lines and redistributing the charge between the charge according to the information in these memory cell capacitances and the charge in the precharged bit line capacitances in an array composed of multiple transistors similar to early DRAM memory cells, in which write and read operations are separated. As an example, it is shown that when three cells are accessed simultaneously, whether the third cell is a "1" or a "0", the result is the sum or product of the information in the remaining two cells. This method realizes only the basic operations, but by taking advantage of the high integration of DRAM, large-scale operations can be allocated and used based on this operation in neural networks.

Another attempt is to use the memory cell shown in Figure 10.8b to represent different bit values by changing the voltage level stored in the capacitor. Multibit cells in DRAM have been attempted to be developed but have not yet been put to practical use. It is hoped that this will be possible with the application of artificial intelligence processing in in-memory computing.

Next, Figure 10.9a shows a method of performing operations as a neural network by setting up logic circuits in units of banks. Eight memory banks are available, and since the calculation function of each bank can be used simultaneously by multiple users, parallel processing is possible, enabling high-speed, large-scale calculations to be performed efficiently.

In this configuration, the logic circuitry is integrated into the memory cell array and used on a bank-by-bank basis. This allows the internal bandwidth of the memory to be maximized. For example, in bank #1, there is a latch circuit that is originally provided at the output end of the memory cell array, and as logic circuits, a multiplier, an adder, and a register are arranged. These perform multibit processing conversion and sum-of-products operations that take advantage of the parallelism of the memory cell array outputs. Matrix operations can be performed in physically close proximity using the information in the memory cell array. Furthermore, there are eight of these, from bank #1 to bank #8 in this figure, operating in parallel. In addition, by controlling the mutual movement of multiple memory banks, parallel processing can be achieved with effectively reduced delays in the back-and-forth of data between banks. At this time, the memory cell array itself does not need to be modified. The memory cell array remains highly integrated even for each bank and can be realized by adding relatively small operations to it. Therefore, it is very powerful for speeding up deep learning operations, which are large data-driven operations.

Figure 10.9b shows a three-dimensional system in which multiple memory chips are placed on the upper layer of a logic chip. It enables energy-efficient data communication between the logic chip and the memory chips, and the two chips placed in close proximity above and below the logic chip cooperate closely to perform data processing. The logic chip contains the computation elements and simultaneously performs multiple calculations, such as matrix and vector calculations. The memory chip stores data, which of course serves as input to the calculations, and the calculation results are also stored in memory. This allows data and calculations to work together

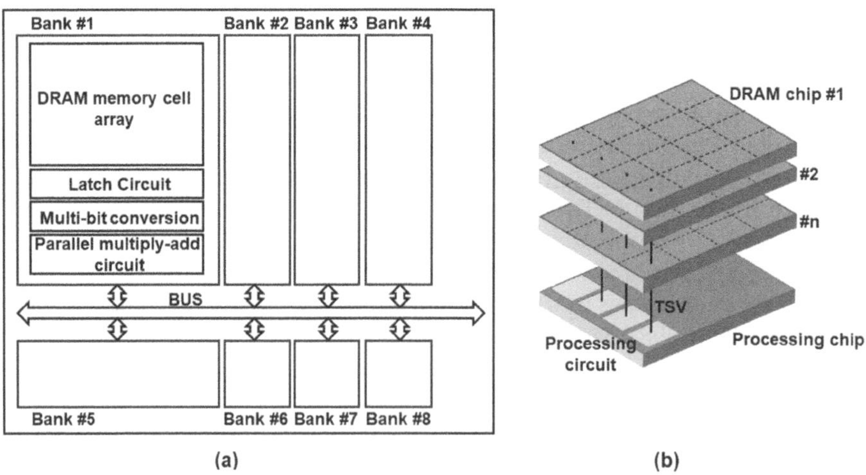

FIGURE 10.9
Other configuration examples using DRAM. (a) Architecture with DRAM arrays (banks) and (b) 3D implementation.

efficiently and enables high-speed calculations. Furthermore, architectural innovations have been made to optimize data communication. This minimizes energy consumption for data movement and transfer while achieving high-performance calculations.

10.3.4 Configuration with RRAM and Non-Volatile RAM

Next, we will show the characteristics of in-memory computing using RRAM and nonvolatile RAM [12,13], which will also be combined with Figure 10.4. Many developments using these are based on edge computing as an application. Due to power supply restrictions, edge computing does not process data all the time, but mostly operates when an event occurs on the spot. Therefore, most of the time when it is used, it is in a standby state. Therefore, the nonvolatility of RRAM and nonvolatile RAM is utilized to not only stop computation but also shut down the power supply. The aim is to build a system with the processing power of about medium required for edge computing. The challenge is that many nonvolatile memories are difficult to manufacture, leaving reliability as an issue. Early practical application will require design within these constraints.

A memory cell array for in-memory computing using RRAM and nonvolatile RAM is shown in Figure 10.10a. This is also a normal memory cell array, which can be used as is. The sum-of-products operation required for neural network processing can be realized by Ohm's law and Kirchhoff's law. In this figure, WL_1–WL_4 are word lines, BL_1–BL_4 are bit lines, and SL_1–SL_4 are source lines. In correspondence with a neural network, the source lines are the inputs, and the bit lines are the outputs that are input to the activation function. Memory

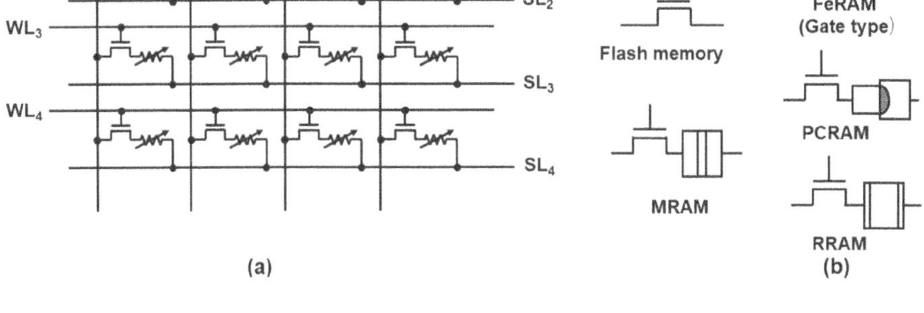

FIGURE 10.10
Computing in memory using RRAM. (a) Memory array configuration and (b) resistor+select device.

cells are placed at these intersections, and the values of the resistors in them correspond to the weights. Depending on how the resistor elements that can represent these weights are configured, memory cells can be incorporated as shown in Figure 10.10b, including flash memory. Among these, flash memory and FeRAM, which have a ferroelectric material placed on the insulating film under the gate, integrate the selection transistor and the resistive element. The resistance changes to represent the weight by changing the threshold of the selection transistor. In other MRAM, PCRAM, and RRAM, the corresponding resistive elements and selection transistors are separated.

Here, in many of the RRAMs and nonvolatile RAMs, the value of this resistance can take multivalued or analog values. In other words, the weights can also be non-binary. This is one reason why RRAM and nonvolatile RAM are increasingly being applied to in-memory computing. As shown in Figure 10.11, if the information stored by this memory cell is one bit, as in other memories, to represent multiple bit weights, multiple word lines must be started up and multiple memory cell information must be transmitted to the bit lines, as shown in (a). On the other hand, if the resistor elements are multivalued or analog values, as shown in (b), only one cell can represent the corresponding multibit or analog value weights.

Figure 10.12 shows the specific processing using this memory cell array. This is exactly the same processing as with the 2D array described so far in this book. First, see Figure 10.4 for the correspondence between the neural network and this array. It is a configuration with three neurons in one layer, where SL_1–SL_3 is the input X_1–X_3 and BL_1–BL_3 is the output Y_1–Y_3, which is also the input to the activation function circuit, not shown in this figure. Figure 10.12a shows the operation in forward propagation used in inference. When all the word lines WL_1–WL_3 are selected, the selection transistors

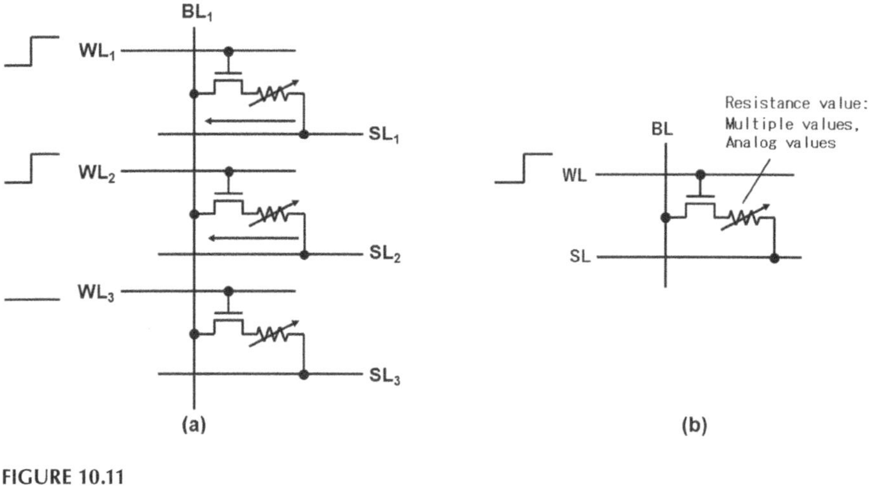

FIGURE 10.11
Realization of multiple bit weights. (a) Multiple cell selection and (b) multivalue or analog value in cell.

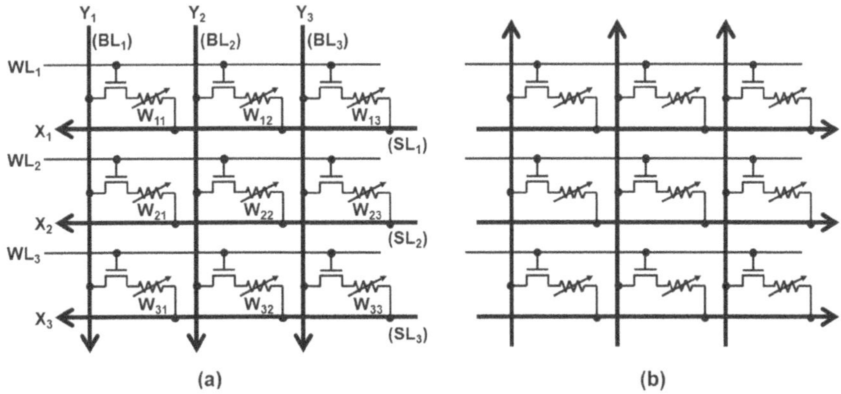

FIGURE 10.12
Forward and backward propagation. (a) Forward propagation and (b) backward propagation.

connected to them are turned on. Let us ignore the resistance of this turned-on selective transistor. The sum-of-products operation required for neural network processing can be realized by Ohm's law and Kirchhoff's law, and let W_{11}–W_{33}, which is the weight shown by each resistor, be expressed here as the conductance, which is the reciprocal of the value of that resistor, and also let input X_1–X_3 be a voltage signal and output Y_1–Y_3 is the current signal, the following equation is obtained. In other words, the sum-of-products operation performed by three neurons is complete.

$$Y_1 = X_1 \times W_{11} + X_2 \times W_{21} + X_3 \times W_{31}$$

$$Y_2 = X_1 \times W_{12} + X_2 \times W_{22} + X_3 \times W_{32}$$

$$Y_3 = X_1 \times W_{13} + X_2 \times W_{23} + X_3 \times W_{33}$$

This $Y1$–$Y3$ is input to the activation function circuit, which is repeated but not shown in this figure.

This configuration can be used as is for learning. As mentioned in Chapter 4, in learning, along with forward propagation, the propagation shown in Figure 10.12b is also used. Here, Y_1–Y_3 is the input and error, and X_1–X_3 is the output. The sum of the products of the weights and errors is calculated, and then the weights are corrected. This calculation is also based on the following equation, which corresponds to the signal flow in this memory cell array configuration.

$$X_1 = Y_1 \times W_{11} + Y_2 \times W_{12} + Y_3 \times W_{13}$$

$$X_2 = Y_1 \times W_{21} + Y_2 \times W_{22} + Y_3 \times W_{23}$$

$$X_3 = Y_1 \times W_{31} + Y_2 \times W_{32} + Y_3 \times W_{33}$$

However, unlike (a), in (b) the sum-of-products operation is realized by Ohm's law and Kirchhoff's law, so Y_1–Y_3 treats the error as a voltage signal and X_1–X_3 as a current signal.

Of course, when using the inverse error propagation method, a circuit to detect this X_1–X_3 signal is necessary, but the sum-of-products operation can be performed only in the circuit shown in Figure 10.12, that is, in the memory cell array of a high-density ReRAM.

However, as mentioned ReRAM technology is not yet mature for manufacturing, these operations are ideal, and in reality, there is significant variation in each resistive element. Therefore, their application is in development.

Furthermore, attempts have been made to incorporate advanced arithmetic operations. There, a single cell may be equipped with multiple resistive elements and multiple selective transistors. These are also used to solve problems with resistive elements, such as variation and erroneous writing during readout. Other arithmetic functions are also incorporated in the sense amplifier section.

In addition to research into configurations such as these with additional operations in the RRAM array and sense amplifier section, it is important to integrate this into a complete computing system. An example of this configuration is shown in Figure 10.13. Here, the CPU, main memory (work memory), nonvolatile memory, and other circuits as accelerators of operations are integrated to realize an advanced neural network.

However, related to what is discussed previously in Figure 10.3 for the system as a whole, it is also important to take the chiplet technology approach of designing multiple small modular chips (chiplets) and combining them to function as a single processor, rather than creating a single large integrated circuit. This approach is also important. Artificial intelligence processing requires high computational power, and chiplet technology allows the use of multiple

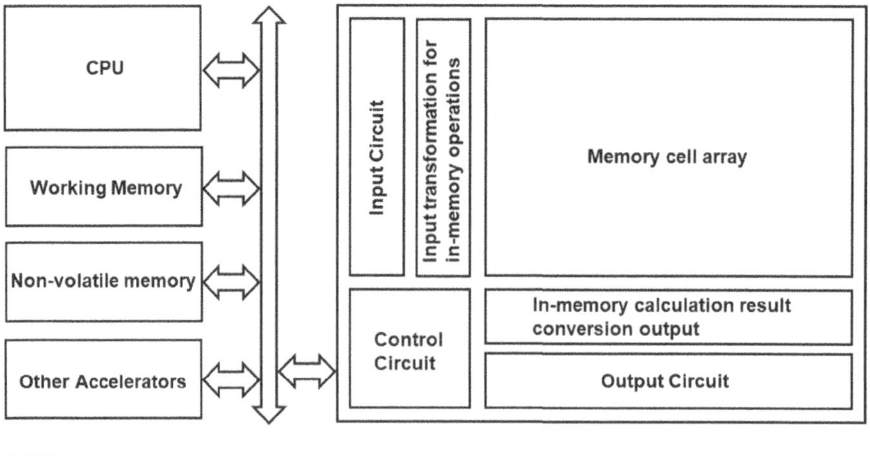

FIGURE 10.13
RRAM coprocessor configuration.

chiplets specialized for different tasks, improving overall performance. It also has the effect of reducing manufacturing costs. While it is complex and expensive to manufacture a large chip at a time, chiplets allow the use of smaller, more efficient silicon die, which improves manufacturing yields. Chiplets can also shorten data travel distances and minimize energy loss and heat generation. Furthermore, there would be increased flexibility to scale up or scale out artificial processing power. Rather than designing a new chip, existing chiplets may be combined to quickly build a processor to meet specific needs.

10.4 Summary

This chapter described "near-memory computing" and "in-memory computing," two promising data-centric signal processing techniques in AI chips. The transfer of data between chips has been shown to have a substantial impact on the performance of the system and its power efficiency. Neural networks are composed of a structural element that performs a sum-of-products operation (MAC operation), utilizing stored weights and the outcomes of the preceding stage's operation. Consequently, the integration of the computing and memory sections, or their proximity, enables the preservation of the neural network's functionality while concurrently achieving a substantial enhancement in the system's power efficiency. In the context of near-memory computing, the arithmetic circuit chip and memory chip are physically positioned in close proximity within the same package or on the same substrate. This configuration is designed to facilitate high-speed, low-power data transfer. Conversely, in-memory computing methodologies implement arithmetic functions within the memory cell array, thereby facilitating simultaneous memory access and arithmetic operations within the same structure. This approach utilizes the inherent characteristics of the two-dimensional memory cell array configuration to ensure efficient processing of matrix operations and other associated tasks.

References

1. D. Patterson, T. Anderson, N. Cardwell, R. Fromm, K. Keeton, C. Kozyrakis, R. Thomas and K. Yelick, "A Case for Intelligent RAM," in *IEEE Micro*, vol. 17, no. 2, pp. 34–44, 1997. https://doi.org/10.1109/40.592312
2. Z. Or-Bach, "A 1,000x Improvement in Computer Systems by Bridging the Processor-Memory Gap," *2017 IEEE SOI-3D-Subthreshold Microelectronics Technology Unified Conference (S3S)*, Burlingame, CA, USA, 2017, pp. 1–4. https://doi.org/10.1109/S3S.2017.8309202

3. G. Singh, L. Chelini, S. Corda, A. J. Awan, S. Stuijk, R. Jordans, H. Corporaal and A. J. Boonstra, "A Review of Near-Memory Computing Architectures: Opportunities and Challenges," *2018 21st Euromicro Conference on Digital System Design (DSD)*, Prague, Czech Republic, 2018, pp. 608–617. https://doi.org/10.1109/DSD.2018.00106
4. M. Aoyagi, F. Imura, F. Kato, K. Kikuchi, N. Watanabe, M. Suzuki, H. Nakagawa, Y. Okada, T. Yokoshima, Y. Yamaji, S. Nemoto, T. T. Bui and S. Melamed, "Developing a Leading Practical Application for 3D IC Chip Stacking Technology," *Synthesiology*, vol. 9, no. 1, pp. 1–15, 2016. https://doi.org/10.5571/syntheng.9.1_1
5. M. Gokhale, S. Lloyd and C. Macaraeg, "Hybrid Memory Cube Performance Characterization on Data-Centric Workloads," *Proceedings of the 5th Workshop on Irregular Applications: Architectures and Algorithms*, Association for Computing Machinery, New York, NY, USA, Article 7, 1–8, 2015. https://doi.org/10.1145/2833179.2833184
6. "Wide I/O Single Data Rate, JEDEC Solid State Technology Association, Standard JESD229", 2011.
7. A. K. Kabat, S. Pandey and V. T. Gopalakrishnan, "Performance Evaluation of High Bandwidth Memory for HPC Workloads," *2022 IEEE 35th International System-on-Chip Conference (SOCC)*, Belfast, United Kingdom, 2022, pp. 1–6. https://doi.org/10.1109/SOCC56010.2022.9908071
8. D. Kim et al., "An Overview of Processing-in-Memory Circuits for Artificial Intelligence and Machine Learning," in *IEEE Journal on Emerging and Selected Topics in Circuits and Systems*, vol. 12, no. 2, pp. 338–353, June 2022. https://doi.org/10.1109/JETCAS.2022.3160455
9. B. Pan, G. Wang, H. Zhang, W. Kang and W. Zhao, "A Mini Tutorial of Processing in Memory: From Principles, Devices to Prototypes," in *IEEE Transactions on Circuits and Systems II: Express Briefs*, vol. 69, no. 7, pp. 3044–3050, July 2022. https://doi.org/10.1109/TCSII.2022.3172494
10. C.-J. Jhang, C.-X. Xue, J.-M. Hung, F.-C. Chang and M.-F. Chang, "Challenges and Trends of SRAM-Based Computing-In-Memory for AI Edge Devices," in *IEEE Transactions on Circuits and Systems I: Regular Papers*, vol. 68, no. 5, pp. 1773–1786, May 2021. https://doi.org/10.1109/TCSI.2021.3064189
11. V. Seshadri, D. Lee, T. Mullins, H. Hassan, A. Boroumand, J. Kim, M. A. Kozuch, O. Mutlu, P. B. Gibbons and T. C. Mowry, "Ambit: In-Memory Accelerator for Bulk Bitwise Operations Using Commodity DRAM Technology," *2017 50th Annual IEEE/ACM International Symposium on Microarchitecture (MICRO)*, Boston, MA, USA, 2017, pp. 273–287. https://doi.org/10.1145/3123939.3124544
12. J.-M. Hung, C.-J. Jhang, P.-C. Wu, Y.-C. Chiu and M.-F. Chang, "Challenges and Trends of Nonvolatile In-Memory-Computation Circuits for AI Edge Devices," *IEEE Open Journal of the Solid-State Circuits Society*, vol. 1, pp. 171–183, 2021. https://doi.org/10.1109/OJSSCS.2021.3123287
13. R. Mochida, R. Mochida, K. Kouno, Y. Hayata, M. Nakayama, T. Ono, H. Suwa, R. Yasuhara, K. Katayama, T. Mikawa and Y. Gohou, "A 4M Synapses integrated Analog ReRAM based 66.5 TOPS/W Neural-Network Processor with Cell Current Controlled Writing and Flexible Network Architecture," *2018 IEEE Symposium on VLSI Technology*, Honolulu, HI, USA, 2018, pp. 175–176. https://doi.org/10.1109/VLSIT.2018.8510676

11

Future Prospects

As we have seen, information processing using artificial intelligence LSIs is not based on the conventional method in which data is exchanged between an arithmetic unit and memory, but rather on a method in which chunks of data are processed together in an arithmetic unit, so to speak.

In the conventional method, in which the arithmetic unit and memory exchange data as they proceed, data are exchanged in parts as the information is processed. In this method, the arithmetic unit is the part that performs calculations and operations, and the memory is a temporary storage place for data. The operator takes one piece of data, processes it, returns the result to memory, and then processes the next piece of data, and so on. On the other hand, the processing often used in artificial intelligence mainly involves sending data in batches to an arithmetic unit and processing them all at once in the arithmetic unit. The chunks of data are processed by the arithmetic unit in batches. This method enables efficient and high-speed processing, and a high power-to-performance ratio can be achieved for complex operations and large amounts of data.

In this chapter, we will look at how these methods evolved and what developments are next, albeit at a conceptual level.

11.1 Development of LSI Technology

The future development of the LSI technology itself that realizes artificial intelligence LSIs can be considered as shown in Figure 11.1, as the time when it will be widely used, although it is possible that it will be developed earlier. It is important to see what technologies can be used in each period, which will determine the type and configuration of artificial intelligence LSIs.

11.1.1 Semiconductor Processing Technology, Device Technology, and New Materials

It is said that values such as 2 and 3nm used when talking about semiconductor technology no longer indicate the minimum processing dimensions. On the other hand, the lattice constant of Si crystal is 0.54nm, and the progress of miniaturization itself will not last long. 3D and (other than power semiconductors)

DOI: 10.1201/9781003599067-11

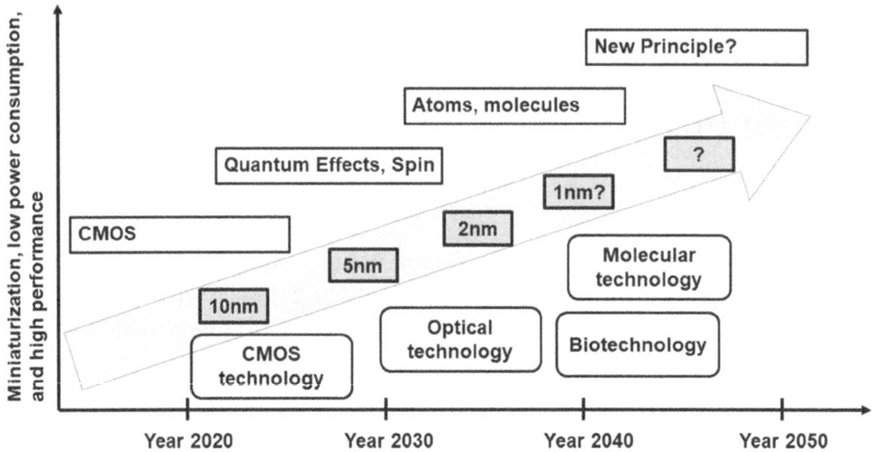

FIGURE 11.1
Development of LSI technology.

new materials are also expected to replace Si. As mentioned in the previous chapter, 3D has already achieved some results, but there are also those who aim for devices using molecular devices and self-assembled materials [1], or for technological innovation in organic materials, biomaterials, or optoelectronic integration [2], rather than in the direction of miniaturization. Also, the development of materials that take advantage of the unique properties of electronic systems that appear on the surface of materials [3] is highly anticipated.

Neural networks and artificial intelligence LSIs themselves were also technologies that aimed to break through the limitations of microfabrication technology. In this context, in-memory computing, which has developed as a high-speed and low-power technology in current processors and memories, is a good match, as we have seen. Low-voltage and low-power circuit technologies have also made significant progress, and their incorporation is expected to continue for the foreseeable future.

From a different perspective, the question of how the manufacture of artificial intelligence LSI can contribute to carbon neutrality as a whole, including product development, will also be raised.

11.1.2 Future Technologies

As already mentioned for some of the quantum devices, quantum computing [4,5] is expected to develop by taking advantage of the quantum size characteristics of Si using microfabrication technology. In particular, since artificial intelligence LSIs are composed of memory and arithmetic operations in an integrated configuration, a configuration that integrates quantum memory and quantum arithmetic operations, which is expected to be developed, will also be developed.

Both optical elements and photoelectric fusion are also areas that await future development as quantum computing [6] using these elements. In the context of quantum computing, the wavelength of light we usually deal with exceeds 10,000°C in terms of temperature, which means that devices using light at room temperature behave in a quantum mechanically stable manner similar to electronic devices including Si at cryogenic temperatures. Large-scale quantum computing at room temperature is expected.

As in the past, development of soft (in the mechanical sense) artificial intelligence hardware using molecular elements and biotechnology, elements with high affinity with living organisms, and artificial intelligence hardware that can be implanted in humans will also continue to progress.

11.2 Development of Artificial Intelligence LSI Systems

As shown in Figure 11.2, the range of applications of artificial intelligence LSIs is expected to expand further as they become faster, more multifunctional, and more low-power. Among these, we summarize the direction of future development of artificial intelligence LSIs and the issues that are considered important, though they have not been discussed much in the previous chapters.

(1) Integration of hardware and software: Optimization through the combination of hardware and software is important to realize fast and efficient artificial intelligence processing. As we have seen, the processing of artificial intelligence LSIs is data-driven. However, when using an in-memory computing chip, for example, the size of data to be processed at a time and the time to access the data differ

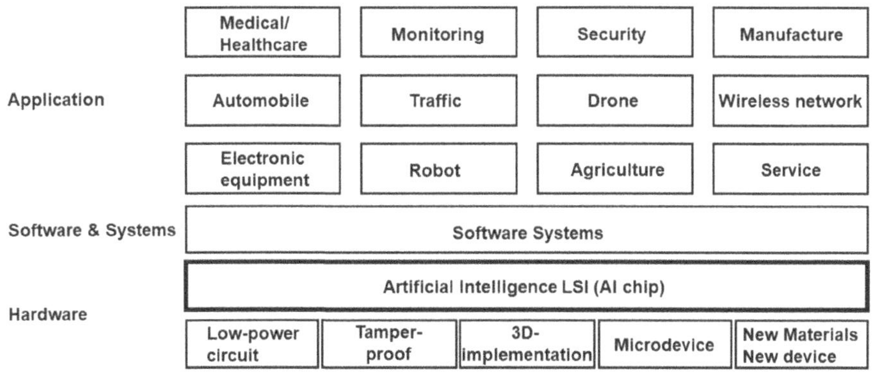

FIGURE 11.2
Artificial intelligence LSI and systems/applications.

depending on whether it is flash memory-based or MRAM-based. It is necessary to take advantage of the hardware characteristics in software so as not to lose the versatility of the application.

(2) Strengthening data security: Artificial intelligence processing LSIs create significant value from data. The value of the data handled in this process is also high. Therefore, along with high processing performance, it is increasingly important to protect the data itself and the value created from it. It is also important to incorporate security-enhancing technologies into the artificial intelligence processing LSI itself, including tamper-resistant technologies [7] that prevent unauthorized analysis and reading from outside.

(3) Integration with robotics: Robotics is an area that requires high-performance specifications and has a large market for artificial intelligence LSIs. The high speed and low power consumption of LSIs are essential for their use in robots. Processing at the edge will include AI control of equipment with moving parts. Various sensors will be attached to the moving parts, and their information will be included. Advanced processing, including control of the moving parts by artificial intelligence LSIs at the edge, will be necessary for autonomous operation and execution of complex tasks. In addition, there are high expectations for new developments in this field as artificial intelligence itself acquires physicality.

11.3 Developments in the Technical Areas of Influence

Deeper involvement with brain science and brain-based computers, which are also the source of artificial intelligence research, is considered necessary when aiming for future dramatic improvements in the development of artificial intelligence LSI technology, especially in the power-to-performance ratio. An example is shown in Figure 11.3. There are some fields where it is difficult to apply current artificial intelligence LSIs as they are, but that is why it is important to develop artificial intelligence LSIs in line with progress in these fields.

(1) Progress in neuromorphic engineering: Neuromorphic engineering [8] refers to the development of hardware that can process information in a manner similar to the brain to create more efficient, adaptive, and intelligent computing systems, and to engineer the brain itself. This is a design concept that differs from the direction of conventional artificial intelligence LSI development as an accelerator

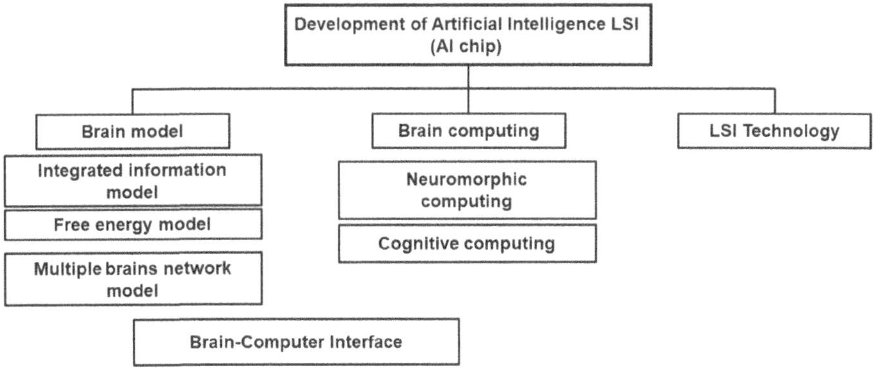

FIGURE 11.3
Technologies supporting the development of artificial intelligence LSI.

for general-purpose information processing. For example, the frequency does not necessarily have to be faster. This may lead to useful developments.

(2) Advances in cognitive computing: Intertwined with neuromorphic engineering, the advancement of research on human cognitive abilities and information processing makes it important to develop algorithms and systems that incorporate new knowledge in this area. Recent advances in LLM are remarkable, and the development of LLM systems is expected to be a major step forward. In the future, however, systems with advanced cognitive functions that are not extensions of the present may be realized for language comprehension, semantic interpretation, and knowledge acquisition, enabling processing similar to intellectual activities such as human. This is an algorithmic advance, but the development of artificial intelligence LSIs suitable for this purpose is desirable.

(3) The brain is thought to understand reality by predicting and integrating external information. The Integrated Information Theory is an attempt to elucidate a method [9] for information processing that mimics such a brain mechanism. Although the role of artificial intelligence LSIs here is unknown, they are expected to contribute to more precisely reproducing the characteristics of the brain's information processing.

(4) Development of the free energy principle for the brain: The brain processes information in an energy-efficient manner. The Free Energy Principle [10] is a model that assumes that the brain as a whole operates on the principle of minimizing the energy required for information processing , thereby explaining the brain's information processing.

Expectations are high, and based on the findings in this field, research on algorithms and circuit design for more efficient energy use will advance, leading to the development of information processing systems with energy efficiency comparable to that of the brain.

(5) The brain has a very complex network structure in which various functions are distributed. Among them, brain functions have coherent units, each of which recognizes concepts and objects. The theory of the A Thousand Brains [11] and other theories have been discussed, which suggest that these network structures are the key features of brain functions. It is necessary to consider artificial intelligence LSIs within the network structure revealed here. This will also lead to the coupling of artificial intelligence LSIs to create new functions for this.

(6) Evolution of brain–computer interfaces: Slightly off the beaten track, brain–computer interfaces (BCIs) [12] are technologies that measure brain activity and link it to machines and computers. Furthermore, it also includes working with the brain. By applying artificial intelligence LSIs here, we will be able to achieve higher precision in real-time signal analysis and accuracy in motor control, and progress in control and information acquisition via brain waves and neural signals.

Can we make a HAL 9000 [13] or an AI for a cat-shaped robot similar to the one depicted in the Japanese anime? The latter AI is probably not cloud-based. That must be more compact in terms of implementation and power when considered as an LSI. The concept of "singularity" [14] was also discussed. These trends, combined with advances in brain-inspired information processing, aim to enable more advanced knowledge acquisition, meaning comprehension, and imitation of cognitive functions, thereby realizing information processing similar to human intellectual activities. Alternatively, in the context of the objective of flying, it is possible that the evolution will be in a different direction from the structure and function of the brain itself, by analogy with the fact that the airplane came to be rather than by imitating the flapping of a bird's wings. For example, consider the recent development of large language models (LLMs). Writing is a product of human intellectual activity. LLMs that have been trained exclusively on the relationships between words from a vast corpus of sentences have begun to manifest a level of intelligence analogous to that of humans. It appears to possess the capacity to emulate the outcomes of human intelligence and intellect. The question of whether a new kind of intelligence will be born in the future, one that does not fly on wings but rather on airplanes, remains to be seen.

Whether the artificial intelligence LSI (chip) as a means of implementation in these circumstances is also an extension of the status quo will be examined as the LSI itself changes.

11.4 Summary

This book has provided an introductory description of the approach, which is primarily based on the wiring logic control method as a basic large-scale integrated circuit (LSI) method for artificial intelligence (AI) processing. This chapter described a high-level overview of the anticipated evolution of this technology and the potential new developments to be expected in the future. First, LSI technology is expected to contribute significantly to the realization of higher performance and new functions in artificial intelligence LSIs through the practical application of new principle elements, such as quantum and optical devices. In future artificial intelligence LSIs, close integration of hardware and software, further improvement of energy efficiency, secure data processing, and linkage with robotics will be important issues. The development of artificial intelligence LSI technology will be influenced by several factors. Among these are the integrated information model, the free energy model, and the thousand-brains network model, which models brain functions. In addition, advances in neuromorphic engineering and cognitive computing will play a role in the development of artificial intelligence LSI technology. Furthermore, the evolution of brain–computer interfaces (BCIs) is another factor that will influence the future direction of this field.

References

1. J.-R. Koo, S.-W. Pyo, J.-H. Kim, H.-K. Lee and Y. K. Kim, "Self Assembled Monolayer for Molecular Electronics," *Japanese Journal of Applied Physics*, vol. 44, no. 1, p. 566. https://doi.org/10.1143/JJAP.44.566
2. S. Hua, E.Divita, S. Yu, B. Peng, C. Roques-Carmes, Z. Su, Z. Chen, Y. Bai, J. Zou, Y. Zhu and Y. Xu, "An Integrated Large-Scale Photonic Accelerator with Ultralow Latency," *Nature*, vol. 640, pp. 361–367, 2025. https://doi.org/10.1038/s41586-025-08786-6
3. J. K. Asbó'th', O. Lá'szló" and A.' Pá'lyi, *A Short Course on Topological Insulators: Band-Structure Topology and Edge States in One and Two Dimensions*, Spinger Nature, 2016.
4. J. K. Gonzalez-Zalba, S. de Franceschi, E. Charbon, T. Meunier, , M. Vinet and A. S. Dzurak, "Scaling Silicon-Based Quantum Computing Using CMOS Technology," *Nature Electronics*, vol. 4, pp. 872–884, 2021. https://doi.org/10.1038/s41928-021-00681-y
5. H. Mizuno, "Quantum Computing from Hype to Game Changer," *2023 IEEE Symposium on VLSI Technology and Circuits (VLSI Technology and Circuits)*, Kyoto, Japan, 2023, pp. 1–4. https://doi.org/10.23919/VLSITechnologyandCir57934.2023.10185419

6. S. Konno, W. Asavanant, F. Hanamura, H. Nagayoshi, K. Fukui, A. Sakaguchi, R. Ide, F. China, M. Yabuno, S. Miki and H. Terai, "Logical States for Fault-Tolerant Quantum Computation with Propagating Light," *Science*, vol. 383, pp. 289–293, 2024. https://doi.org/10.1126/science.adk7560
7. M. Tehranipoor and C. Wang, Eds., Introduction to Hardware Security and Trust, New York, NY: Springer, 2012.
8. G. Quaglio, P. Toia, E. I. Moser, T. Karapiperis, K. Amunts, S. Okabe, M. M. Poo, J. C. Rah, Y. De Koninck, J. Ngai and L. Richards, "The International Brain Initiative: Enabling Collaborative Science," *The Lancet Neurology*, vol. 20, no. 12, pp. 985–986, 2021. https://doi.org/10.1016/S1474-4422(21)00389-6
9. Tononi and M. Massimini, *The Consciousness Instinct: Unraveling the Mystery of What Makes Us Human*, New York, NY, USA: Penguin Press, 2018
10. T. Parr, G. Pezzulo and K. J. Friston, *Active Inference: The Free Energy Principle in Mind, Brain, and Behavior*, Cambridge, MA: MIT Press, 2022.
11. J. Hawkins, *A Thousand Brains: A New Theory of Intelligence*, New York: Basic Books, 2021.
12. J. Peksa and M. Dmytro, "State-of-the-Art on Brain-Computer Interface Technology," *Sensors*, vol. 23, no. 13, p. 6001, 2023. https://doi.org/10.3390/s23136001
13. A. C. Clarke, 2001: *A Space Odyssey*, New York: New American Library, 1968.
14. R. Kurzweil, *The Singularity Is Near: When Humans Transcend Biology*, New York: Viking, 2005.

Index

Note: **Bold** page numbers refer to tables and *italic* page numbers refer to figures.

M